Mobility

Theory and Practice

Werner Schneider, M.D.
Kreuzlingen, Switzerland

Hans Spring, M.D.
Leukerbad, Switzerland

Thomas Tritschler
Schaffhausen, Switzerland

Foreword by H. Baumgartner, M.D.
Zurich, Switzerland

77 illustrations

1992
Georg Thieme Verlag
Stuttgart · New York

Thieme Medical Publishers, Inc.,
New York

Werner Schneider, M.D.
Hauptstrasse 39
CH-8280 Kreuzlingen

Hans Spring, M.D.
Rheuma- und Rehabilitationsklinik
CH-3954 Leukerbad

Thomas Tritschler
Director, Physiotherapieschule,
Kantonsspital
CH-8208 Schaffhausen

Drawings:
Peter Freitag, Günther Biste

Library of Congress Cataloging-in-Publication Data

Schneider, W. (Werner), 1941–
[Beweglichkeit. English]
Mobility : theory and practice / Werner Schneider, Hans Spring, Thomas Tritschler ; foreword by H. Baumgartner.
p. cm.
Translation of: Beweglichkeit. 1989.
Includes bibliographical references and index.
1. Joints—Range of motion. 2. Biomechanics. 3. Physical therapy. I. Spring, Hans. II. Tritschler, Thomas. III. Title.
[DNLM: 1. Movement. WE 103 S3583b]
QP303.S2413 1992
DNLM/DLC
for Library of Congress 92–403

Translation provided by F. Hoffmann-La Roche Ltd, Basel, Switzerland

This book is an authorized translation of the German edition published and copyrighted 1989 by Georg Thieme Verlag, Stuttgart, Germany. Title of the German edition: Beweglichkeit: Theorie und Praxis.

Thieme Medical Publishers, Inc., 381 Park Avenue South, New York, N.Y. 10016
Typesetting by Druckhaus Götz GmbH, Ludwigsburg (Linotron 202)
Printed in Germany by Druckhaus Götz GmbH, Ludwigsburg

ISBN 3-13-767301-1 (GTV, Stuttgart)
ISBN 0-86577-435-8 (TMP, New York) 1 2 3 4 5 6

Foreword

"Life is movement" is an oft-quoted phrase, but no less true for that. In all its microscopic and macroscopic forms, in the physical and the chemical domains, life means movement. Our civilized way of life demands that most children's spontaneous, aimless urge to move be restrained in a planned and organized manner when they start school in order to make learning possible, with the result that a few years later the school doctor finds a surprisingly high proportion of children in the intermediate classes with marked impairment of mobility due to shortening of the muscles, e.g., the flexors of the hip. In the higher grades, where increasing periods of sitting and a decreasing amount of athletics are the rule, the situation is even worse. Since the late 1940's the traditional school desks with their angled tops have been replaced by horizontal ones. But we cannot sit up straight unless the reading or writing material in front of us is inclined at an angle of 16°. The horizontal surface forces the body into a stressful, bent position and promotes an unhealthy posture. Segments of the spine become stiff; the muscles of the back are strained and under tension, giving rise to pain which may radiate into the arms and legs.

The notion of movement as an expression of life is neglected not only during these formative years but also during the working life of a large segment of the population.Thus, patients with pain of the motor apparatus are increasingly filling doctors' offices from an early age on. Surprisingly although the interest in sports has been growing in recent decades, this development has not changed very much. Admittedly, physical education classes several times a week have become an intrinsic part of the school curriculum, sports camps and swimming pools are available to counteract static stress, and numerous sports clubs offer a variety of activities. And yet the number of people complaining of pain in the motor apparatus does not get any smaller. The examining physician finds tight, shortened muscles and ankylosed spinal segments in a surprisingly high perecentage of cases, even among patients who regularly play sports. Must we conclude that the so numerous sports facilities are still not being used enough? Or are these facilities themselves not suitable to counteract acquired, painful stiffness? It becomes obvious very rapidly that the answer is not to be found in aerobics and indeed that not every part of the body that can be stretched ought to be stretched. Movement is made up of three components: mobility, strength, and endurance. Only when sufficient attention is paid to each of them and physical effort is dosed in accordance with correct energetic principles is one able to resist day-to-day physical stresses.

Thus, it seems highly appropriate that an experienced team composed of a rheumatologist, a sports physician, and a physical therapist should provide practicing physicians and therir patients with a manual that presents basic principles as well as the means of evaluation and therapy of motor disorders. And it is logical, too, that the first volume should deal with mobility as the basis of all movement. I welcome this series and hope that it will enable its readers to instruct and carry out proper movement in line with our present-day understanding.

Spring 1992

H. Baumgartner, M. D.
Chief Physician
Wilhelm Schulthess Clinic
Zurich, Switzerland

Preface

Mobility, besides strength and endurance, is one of the basic locomotive forces that are of great importance not only for efficient physical performance but also for the feeling of well-being.

We worked out the organization of this Flexibook based on our knowledge of manual therapy, physiotherapy, rehabilitation, and sports medicine. Our concept of how to assess and improve upon mobility aims at assisting physicians, physiotherapists, trainers, and sports teachers in their work.

The exercises presented here are suitable for use in the rehabilitation of a wide variety of diseases and injuries, improving mobility with a minimum of strain.

We would like to thank our colleagues B. Schneider, R. van De Felde, I. Kolb, and T. Terwiel for their critical perusal of the manuscript, and P. Freitag for his excellent graphics. Achim Menge of Thieme Verlag and M. Hälg-Zürcher and F. Brunner of F. Hoffmann-La Roche Ltd generously helped us in matters of production and printing. We also received support in didactic and organizational aspects from A. Cueni and U. Lieberherr.

Spring 1992 *W. Schneider, H. Spring, T. Tritschler*

Contents

Introduction

Therapeutic concepts in physical medicine and rehabilitation can be successful only if the patients are motivated to make a substantial contribution themselves: in addition to supervised therapy, they must be willing to carry on alone with specific training of the basic elements of motor function – *mobility, strength, and endurance* – and to continue these steadfastly with prophylactic exercises to prevent recurrences.

Instruction in such a home program must be provided by the attending physician or physical or medical therapist. Its range and intensity must be adapted to each individual clinical picture.

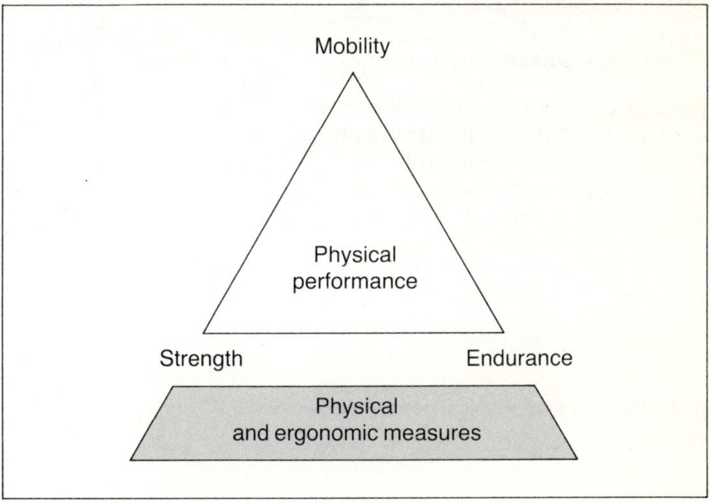

Fig. **1** Structure of the four-part program

Theoretical Principles

Mobility

Mobility is the capacity to perform movements over a large range of motion. It is one of the five basic elements that make up a person's physical performance.

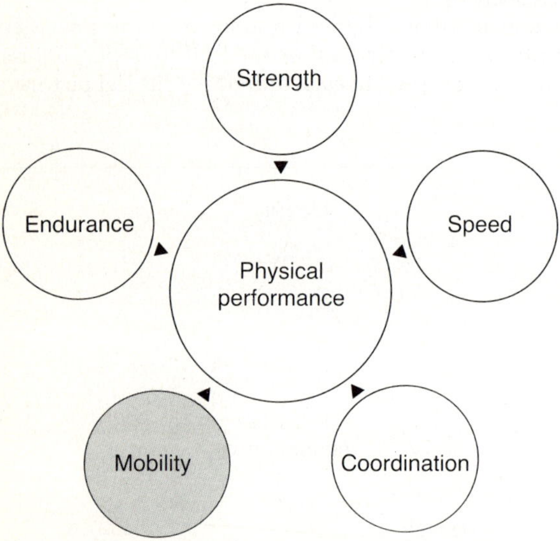

Fig. **2** Mobility and physical performance

A wide variety of factors have a direct influence on mobility:

- Age
- Sex
- Tissue type
- Degenerative and inflammatory spinal and joint disease
- Neurological status
- Congenital and acquired deformities
- Level of fitness
- Mental state
- Temperature
- Time of day

Mobility can be divided into two components: *flexibility* and *extensibility*. Flexibility is a characteristic of the joints and disks; extensibility, a characteristic of muscles, tendons, ligaments, and joint capsules.

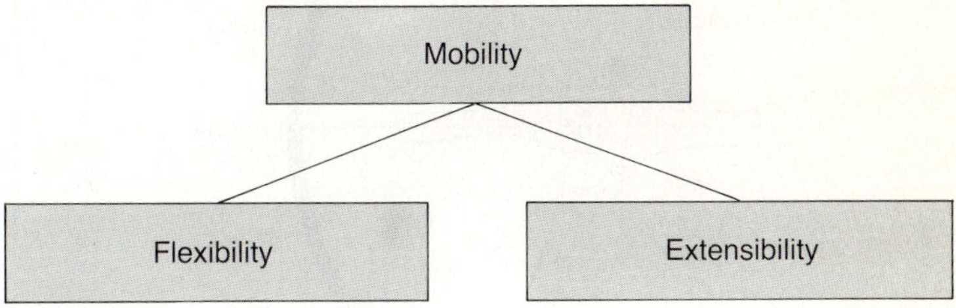

Flexibility

Angular Mobility

Angular movements may be active or passive; they are generally classified as:

- Flexion–extension
- Inclination–reclination
- Lateral flexion
- Abduction–adduction
- Rotation
- Elevation–depression

These angular movements are tested and measured in the classic orthopedic examination (Debrunner: Orthopedic Diagnosis).

Range of Motion

In each plane of movement, joints and spinal segments exhibit different limits of mobility, depending on whether they are moved actively or passively. *Physiological limits of mobility* are limits attainable by active movement. *Anatomical limits of mobility* are limits attainable by passive movement, i.e., through the application of force to a joint.

Normal Mobility

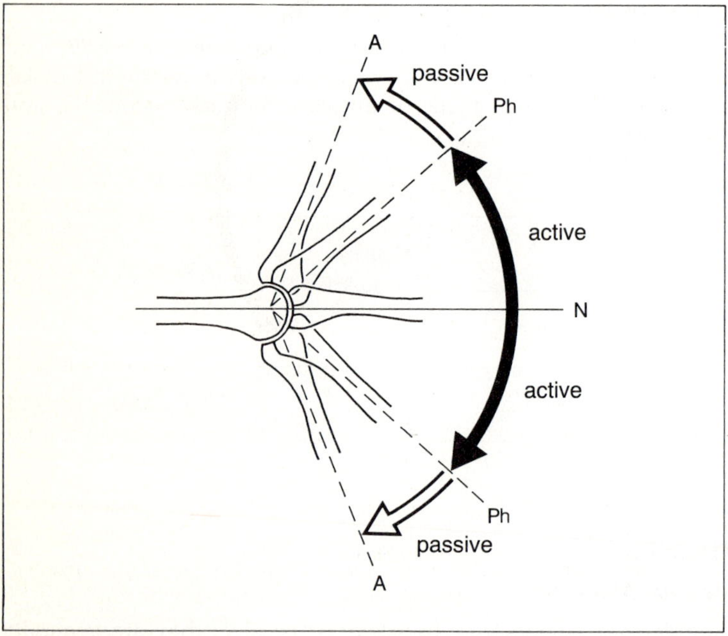

Fig. **3** Normal mobility

A: Anatomical limit of mobility: attained by passive movement (➡)
Ph: Physiological limit of mobility: attained by active movement (⇨)
N: Neutral position

Pathological Mobility

Joint movements may either exceed or fall short of normal limits. Reduced joint mobility is referred to as *hypomobility;* excessive joint mobility, as *hypermobility*.

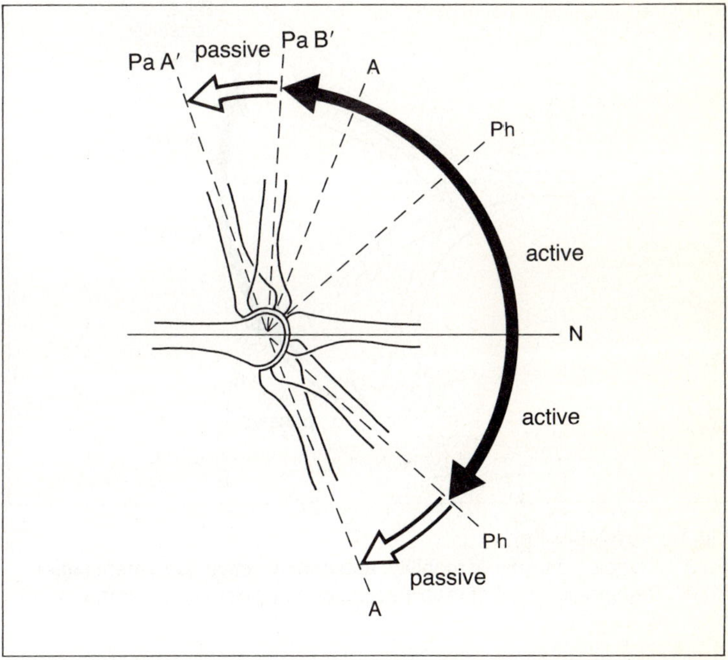

Fig. **4** Hypermobility

Pa B′: Pathological limit of mobility: attained by active movement (➡)

Pa A′: Pathological limit of mobility: attained by passive movement (⇨)

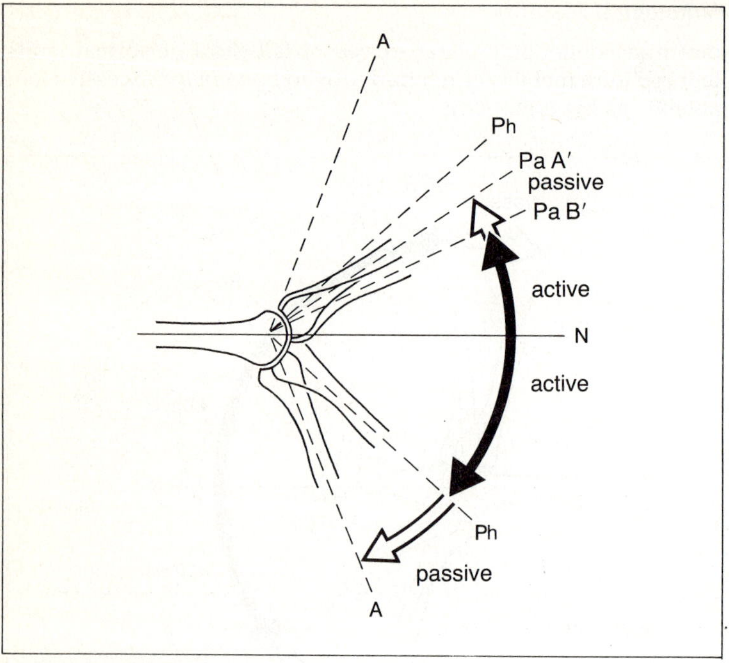

Fig. **5** Hypomobility

Pa B′: Pathological limit of mobility: attained by active movement (➡)
Pa A′: Pathological limit of mobility: attained by passive movement (⇨)

A joint or spinal segment is said to be unstable when it is inadequately stabilized by the ligaments and/or muscles when strain is put on it. *Instability* may manifest itself inside or outside the physiological or anatomical range of movement.

In the vast majority of cases, prolonged instability gives rise to premature degenerative changes in the joints, intervertebral disks, ligaments, tendons, or muscles.

We speak of *hyperlaxity* when constitutional excessive elasticity of the skin, tendons, ligaments, and joint capsules gives rise to hypermobility. There is no hard-and-fast dividing line between normal elasticity and a true Ehlers-Danlos syndrome.

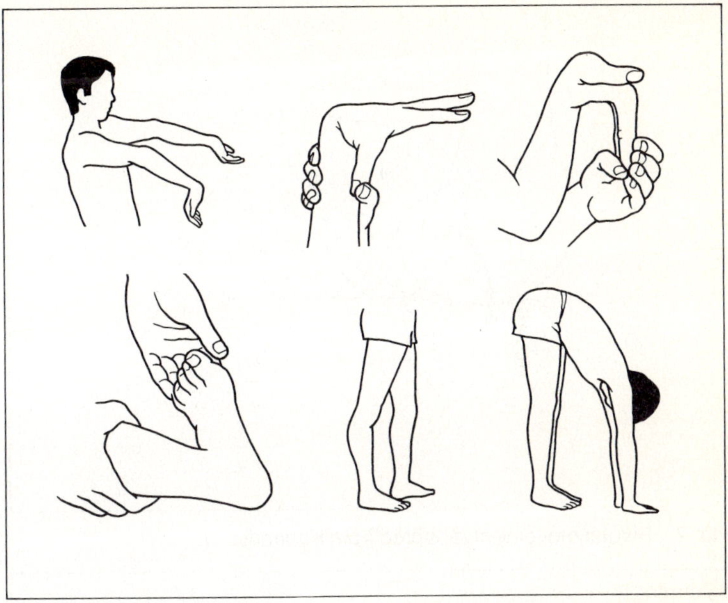

Fig. **6** Hyperlaxity syndrome

The Pivotal Glide

Analysis of what happens inside joints shows that movements always occur in the form of a *pivotal glide.* The architecture of a joint or spinal segment and the arrangement of the ligaments and muscles determine the direction and range of this pivotal glide.

In addition to optimum muscular control, harmonious movement depends to a great extent on the pivotal glide inside the joint or spinal segment. An increase in glide resistance as a result of degenerative or inflammatory cartilaginous changes can impair pivotal glide considerably. Adhesion of joint capsules and tendons or contraction of a joint capsule can render a normal pivotal glide impossible. Muscular imbalance routinely impairs pivotal glide by depriving joints and spinal segments of correct muscular control.

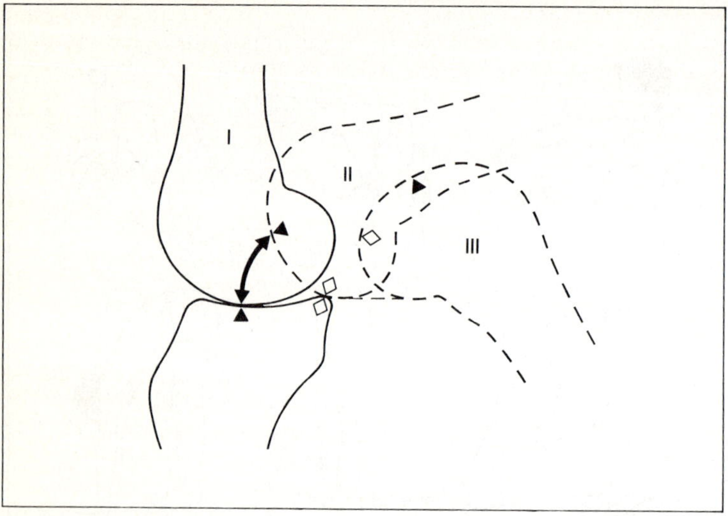

Fig. **7** Pivotal movement (adapted from Kapandji)

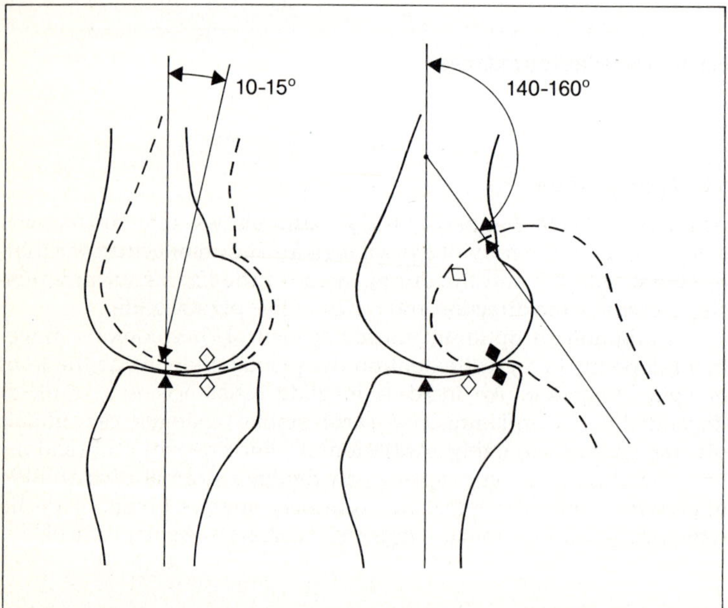

Fig. **8** Pivotal glide (adapted from Kapandji)

When the *distal surface of a joint is concave* its pivotal glide occurs in the same direction as the angular movement; in the case of a *convex distal articular surface,* it occurs in the reverse direction.

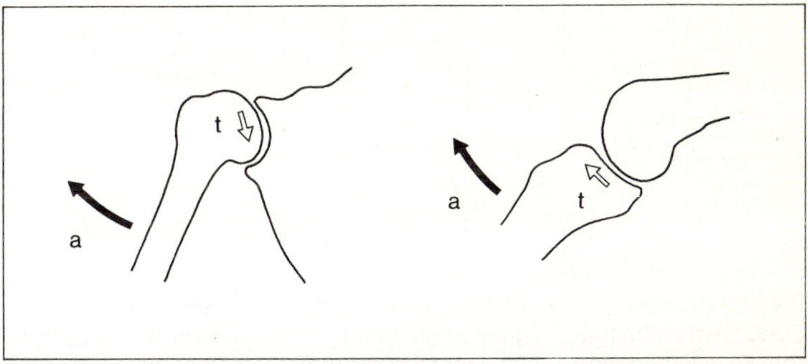

A B

Fig. **9** A Convex rule; B concave rule
a = angular movement (➡); t = translational movement (⇨)

The convex rule
applies to joints whose distal parts have a convex articular surface. It states that translational motion between the articular surfaces will be opposite to the direction of angular motion.

The concave rule
applies to joints whose distal parts have a concave articular surface. It states that translational and angular motion between the articular surfaces will be in the same direction.

Degenerative changes in the cartilage increase the *friction* between the opposing surfaces. The coefficient of friction is also determined by the viscosity of the synovial fluid. The thinner (i.e., the more inflammatory) the synovial fluid, the higher the coefficient of friction.

The forces that produce the gliding component of a pivotal glide increase in proportion to the friction. Passage from the pivotal to the gliding movement takes place in a jerky, "stop–start" fashion. Such an arthrotically altered pivotal glide can be palpated or auscultated as a joint sound.

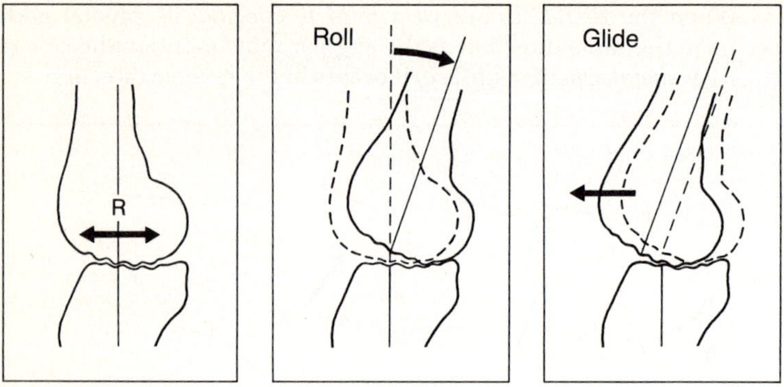

Starting position
Coefficient of the friction (R) is increased by cartilaginous changes

Phase 1 (start of angular movement)
As a result of the high coefficient of friction, the movement is purely pivotal

Phase 2
Continuation of angular movement produces a jerky, rapid gliding movement

Fig. **10**

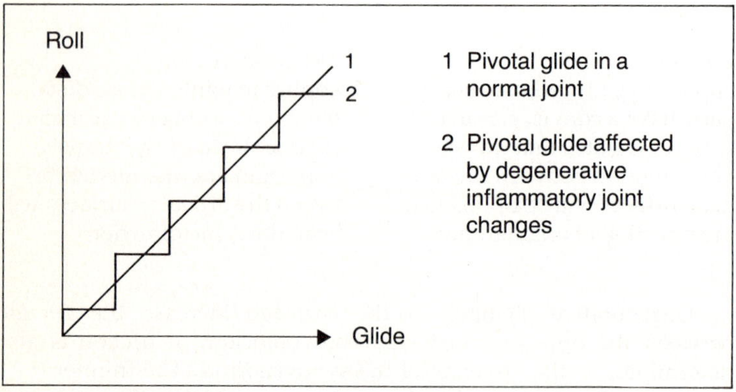

Fig. **11** Joint sounds

Translational Mobility

Translational mobility, like angular mobility, can be tested and treated in any joint or spinal segment.

Every joint movement has a specific translational component. A very close relationship exists between the degree of angular and the degree of translational mobility of joints.

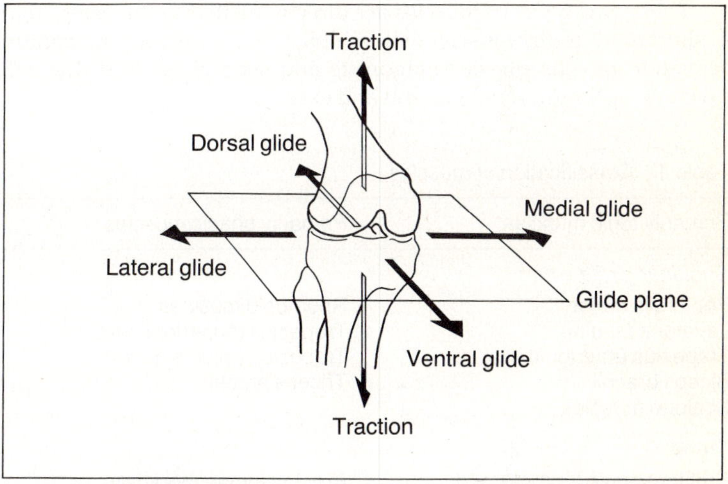

Fig. **12** Translational joint movements

Extensibility

The muscles are another essential element of the motor apparatus; and, like joints, they are treatable.

When muscles react to pathological stimuli, the result is either a structural change (e.g., myogelosis) or muscular imbalance.

Muscular Imbalance

The muscles can be classified phylogenetically as

- tonic,
- mixed,
- phasic.

The tonic muscles originally had a purely postural function, and the phasic muscles figured primarily in movement. Muscle groups with both these functions are termed "mixed."

In humans, tonic and phasic muscles are no longer found in their pure forms. These terms can still be applied to certain muscles, however, because of the ways in which they react to incorrect or excessive loading. Predominantly tonic muscles react by becoming shorter, and predominantly phasic muscles are weakened (Table **1**).

These two types of muscles are directly related to the extent that a shortened tonic muscle will inhibit – i.e., prevent maximum activation of – its phasic antagonists and synergists. And this will preclude optimum rehabilitation and exercise.

Table 1 Classification of muscles

Primarily tonic muscles	Primarily phasic muscles
Shoulder girdle – arm	
Pectoralis major	Rhomboid muscles
Levator scapulae	Trapezius (superior fibers)
Trapezius (inferior fibers)	Trapezius (middle fibers)
Biceps brachii	Triceps brachii
Scalene muscles	
Trunk	
Erector spinae lumbalis and cervicalis	Erector spinae thoracalis Abdominal muscles
Quadratus lumborum	
Pelvis – thigh	
Biceps femoris	Vastus medialis
Semitendinosus	Vastus lateralis
Semimembranosus	Gluteus medius
Iliopsoas	Gluteus maximus
Rectus femoris	Gluteus minimus
Adductor longus	
Adductor brevis	
Adductor magnus	
Gracilis	
Piriformis	
Tensor fasciae latae	
Lower leg – foot	
Gastrocnemius	Tibialis anterior
Soleus	Peroneal muscles

Muscular imbalance is a state of disequilibrium between the tonic and the phasic musculature. The tonic muscles, while retaining their strength, are shortened, whereas the phasic antagonists and synergists retain their normal length but are weakened.

Muscular imbalance may result from a variety of causes:

- Incorrect or excessive loading of the motor apparatus
- Vertebral syndromes
- Spondylogenic syndromes
- Radicular syndromes
- Arthrosis
- Muscular injury
- Joint instability
- Arthritis and spondylitis
- Tendinopathy
- Inactivity and prolonged immobilization
- Psychogenic disorders of the motor apparatus

Muscular imbalance reduces the loading capacity of the motor apparatus. Shortened tonic muscles react more rapidly than the phasic muscles in protective and defensive movements. This can lead to moments of mechanical overloading, with a risk of muscle pulls or even rupture of a shortened tonic muscle. Shortened muscles are less supple and elastic in the relaxation phase. This increased internal resistance is commonly the cause of painful straining of the corresponding muscles and ligaments (e.g., insertion tendinosis). Muscular imbalance disturbs pivotal glide and thus prevents optimum movement of joints and spinal segments.

As a result of regional muscular imbalance, the load placed on joints and spinal segments is increased, and they may become irritated. For example, the iliopsoas, rectus femoris, and erector spinae lumbalis, which are often found to be shortened, together with a weakened gluteus maximus and weak abdominal muscles, cause the pelvis to tip forward and thus produce an orthostatically unfavorable hyperlordosis of the lumbar spine.

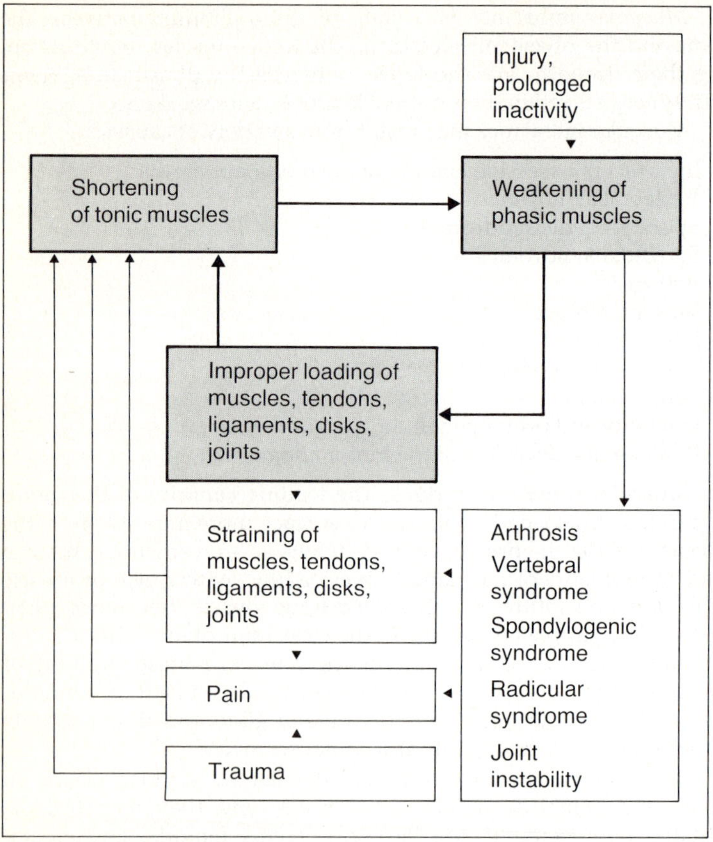

Fig. **13** Causative mechanisms of muscular imbalance

Structural Degenerative Muscle Changes

In addition to muscular imbalance, the skeletal musculature reacts to diseases of the motor apparatus by undergoing structural changes. Electron microscopy, light microscopy and clinical examination show evidence of myogelosis. This is closely related to the "trigger points" of the muscles (Fassbender: *Rheumapathologie;* Simon and Travel: *Trigger Points).*

Myogelosis can affect both phasic and tonic muscles. Muscular imbalance is often observed concurrently.

These degenerative rheumatic myopathies are beyond the scope of this book. The reader should refer to the specialized literature.

Examination Technique

Functional Joint Testing

Angular Movements

Active Movements

In order to obtain reproducible results, patients are to actively perform standardized movements of the joints and spine. Active movements are evaluated in terms of

- range of movement,
- harmony of movement,
- pain during movement,
- pain after movement, and
- strength and pain during movement against resistance.

Any abnormal movement pattern found is further analyzed by passive-movement studies. If necessary, the tests should be supplemented by neurological studies.

Passive Movements

Passive movements are performed along the standard paths of motion (see pp. 19–23). The following criteria are used for evaluation:

- Range of motion
- Type of stop at the limit of movement
- Friction sounds (see p. 10)
- Pain during movement
- Pain after movement

These tests are to be supplemented as necessary by traditional orthopedic, traumatological, rheumatological, and neurological examinations.

Translational Mobility

The joint or spinal segment to be tested is placed in the resting position and should be grasped close to the joint space to prevent angular movement. As translational movements in the glide plane are performed, mild traction is applied (Schneider et al.:
Manual Medicine – Therapy).

The sum of the components of translational mobility of a joint is termed "joint play."

The examination technique for translational mobility is taught today in manual therapy courses and cannot be described in greater detail here.

Types of Stop at the Limit of Movement

The end point of a passive movement provides considerable information on the causes of pathologically reduced or increased mobility (see pp. 4–6). The stop at the limit of movement should be investigated for both angular and translational movement.

The stop at a limit of movement is said to be *hard* when the movement terminates abruptly against *bony* resistance (e.g., an osteophyte).

It is *hard and reflex* when movement is limited by *nociceptive* muscle contractions of very rapid onset (a true Lasègue sign, for instance).

It is *soft* when movement is countered by gradual *tensing of muscles,* ligaments, and joint capsules (e.g., a shortened tonic muscle, joint effusion).

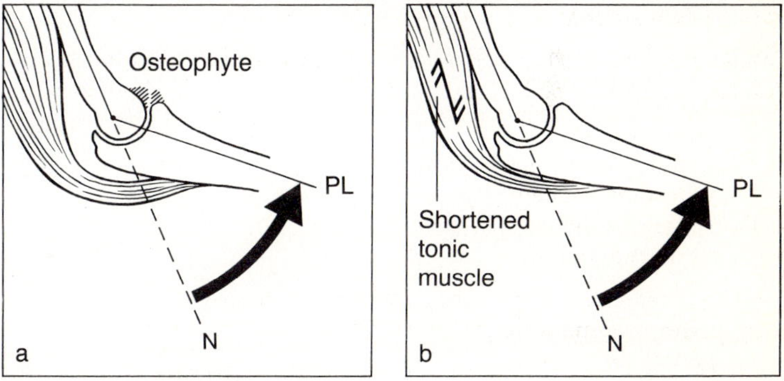

Fig. **14** **a** Hard stop; **b** Soft stop
PL = Pathological limit of movement

If during a joint examination *local and/or referred pain* occurs in addition to a pathological stop, evaluation of the stop often provides a functional as well as a structural interpretation of the pain. This must often be further investigated by imaging and other exploratory techniques.

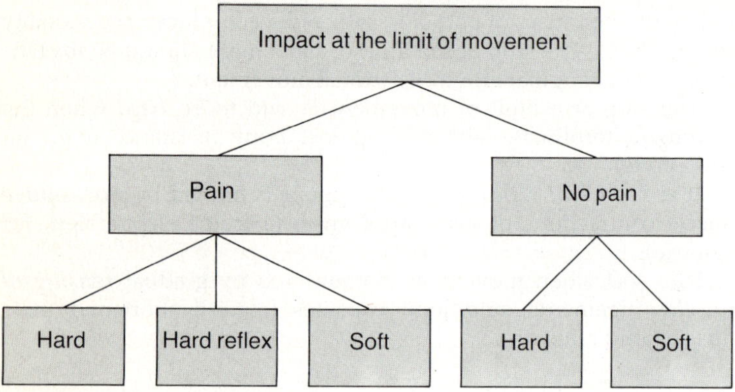

Fig. **15**

Clinical Testing of the Angular Range of Movement

1. Vertebral Column (Fig. **16**)

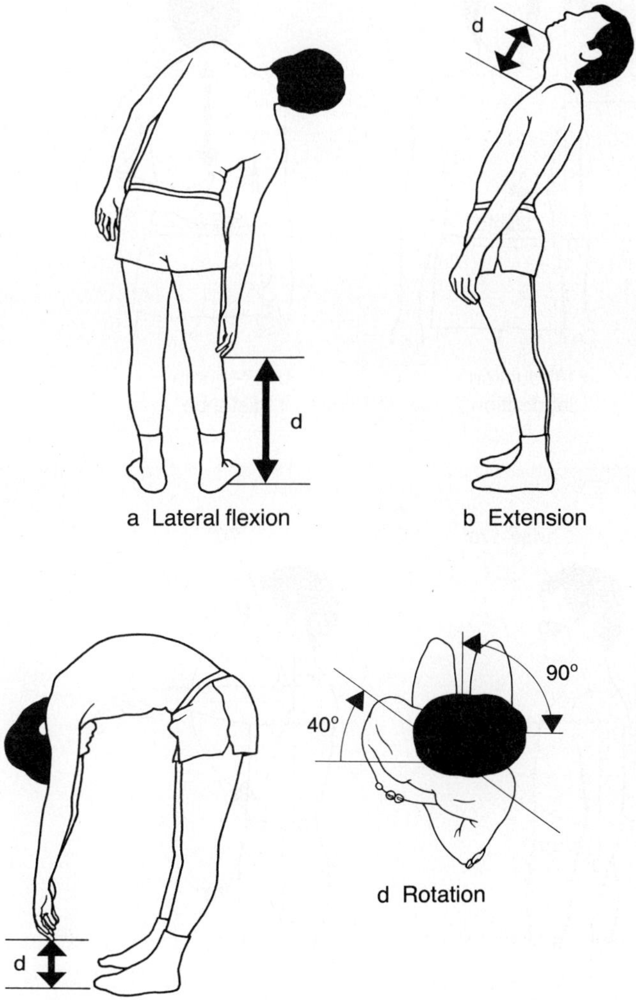

a Lateral flexion

b Extension

c Flexion

d Rotation

Fig. **16**

2. Shoulder Girdle (Fig. **17**)

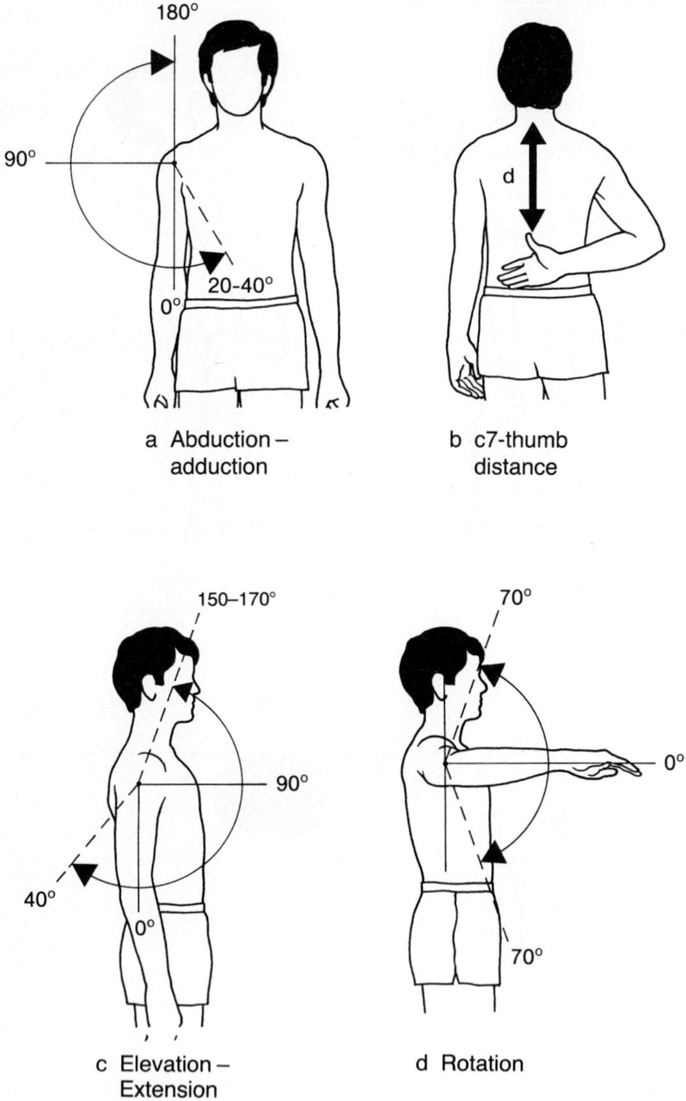

a Abduction – adduction

b c7-thumb distance

c Elevation – Extension

d Rotation

Fig. **17**

3. Elbow and Forearm Joints (Fig. 18)

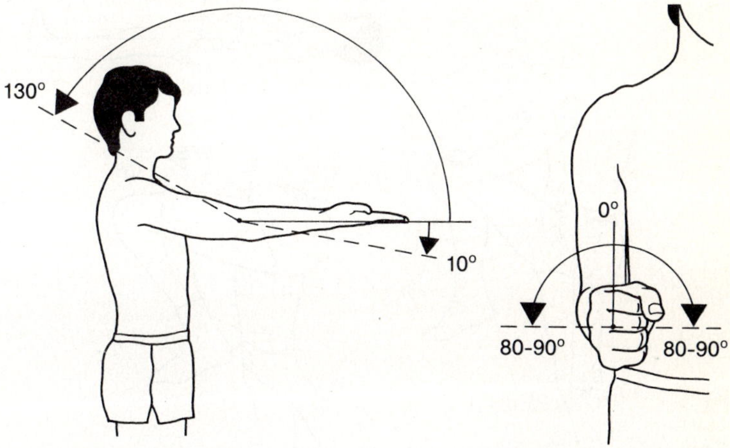

a Flexion – Extension

b Supination – Pronation

Fig. **18**

4. Hip Joint (Fig. **19**)

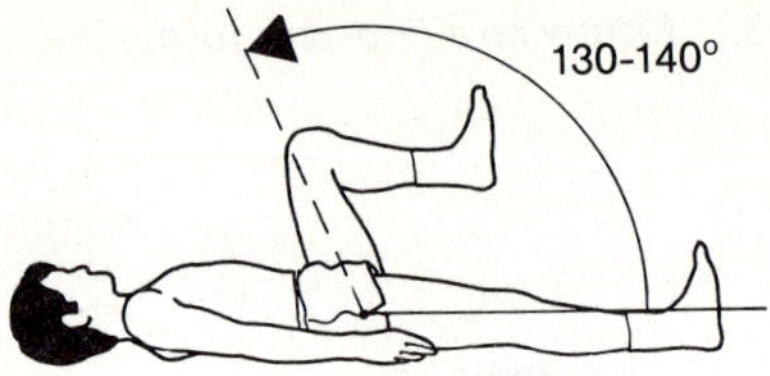

a Flexion

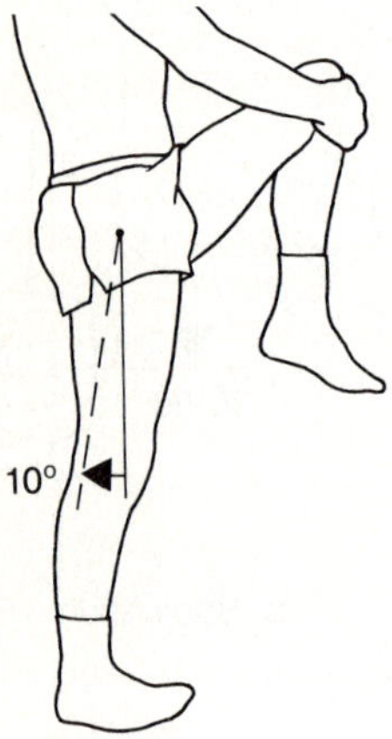

b Extension

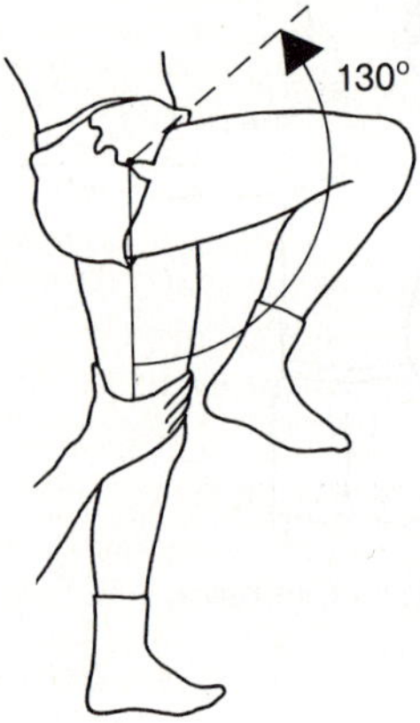

c Flexion

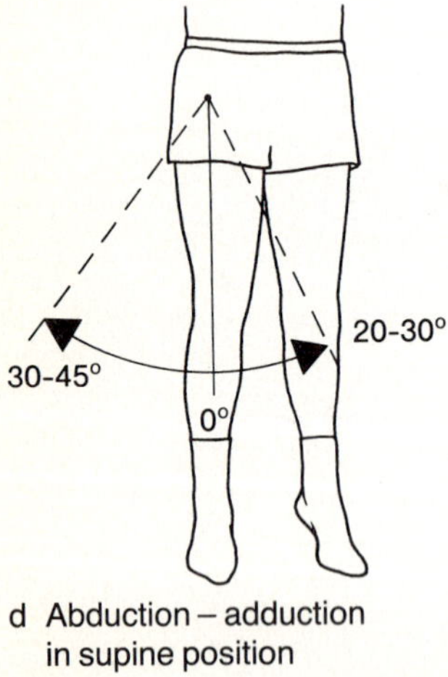

d Abduction – adduction in supine position

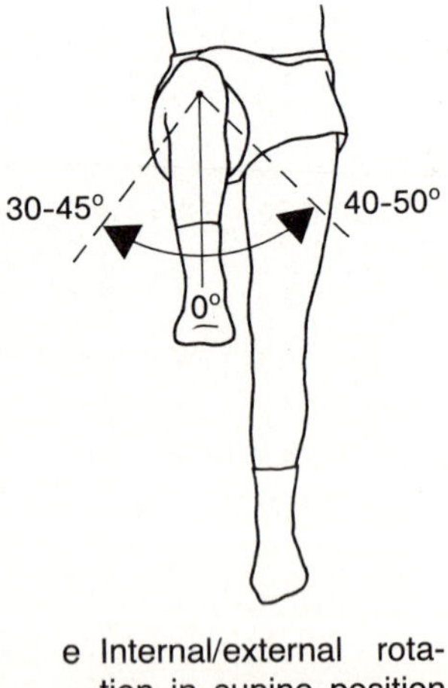

e Internal/external rotation in supine position with hip joint bent

Fig. **19**

5. ***Knee Joint*** (Fig. 20)

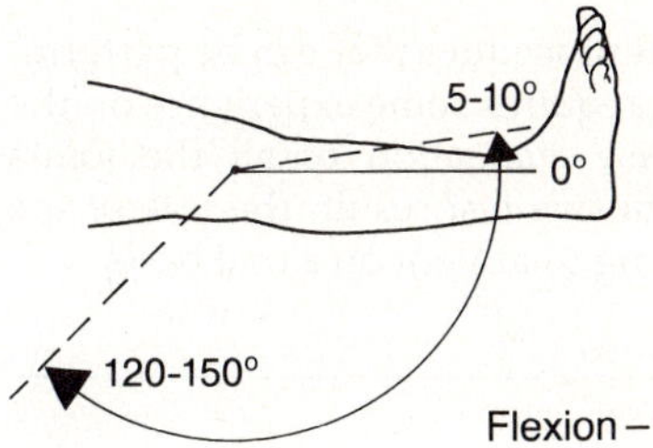

Flexion – Extension

Fig. **20**

6. ***Ankle Joint*** (Fig. 21)

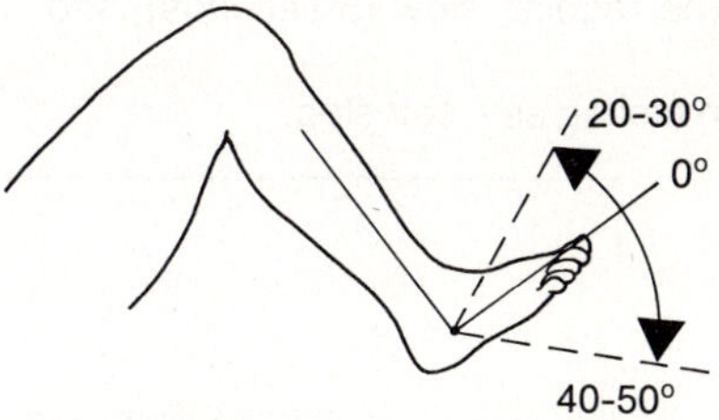

Plantar flexion – Dorsal extension of the ankle joint (adapted from Debrunner and Kapandji)

Fig. **21**

Functional Muscle Testing

Testing of Muscle Length

Testing of muscle length is a clinical procedure that can be performed without the use of equipment. It requires some experience on the part of the examiner and a precise evaluation of all the joints involved. In the event of an inconclusive test result, the role of the muscle should be clarified by initiating treatment on a trial basis.

Principles of Length Testing

1. First carefully examine the joints involved (joint play). Limited joint mobility makes testing more difficult.
2. Examination should center on *one* joint at a time. In the case of biarticular muscles, the other joint should be immobilized.
3. Take the correct starting position.
4. The tested muscle must not be stimulated before and during testing. Your grasp should be as light as possible.
5. Testing is passive, with the patient as relaxed as possible.
6. The test movement should be regular, slow (5 seconds), and fluid.
7. A shortened muscle is felt on testing as a soft stop.

Clinical Testing of Muscle Length

1. Trapezius
2. Levator scapulae
3. Scalene muscles
4. Pectoralis major
5. Extensores carpi et digitorum
6. Erector spinae lumbalis
7. Psoas major
8. Rectus femoris
9. Piriform muscle
10. Adductores femoris
11. Tensor fasciae latae
12. Ischiocrural muscles
13. Triceps surae

1. Trapezius

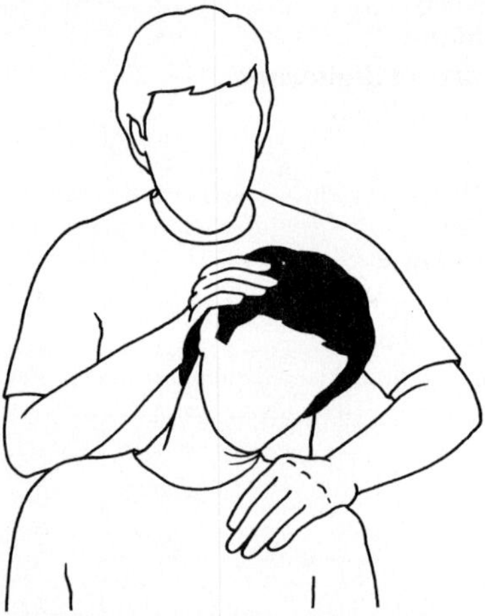

Fig. **22**

Execution Lateral flexion of the cervical spine with slight rotation in the same direction.

Interpretation Premature elevation of the shoulder girdle indicates a shortening of the trapezius (superior fibers).

2. Levator scapulae

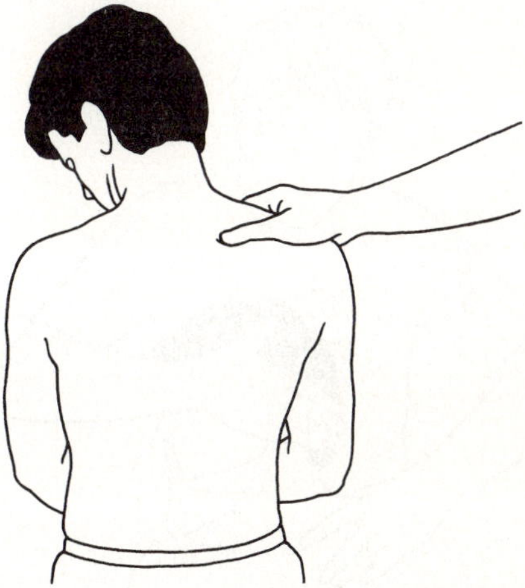

Fig. **23**

Execution	Passive flexion of the cervical spine with rotation to the opposite side. Press simultaneously on the muscle in the region of its insertion.
Interpretation	A burning pain along the course of the muscle indicates shortening. A shortened muscle will react to friction in the region of its insertion with a rapid increase in tone. Crepitation of the paramuscular connective tissues may occur.

3. Scalene Muscles

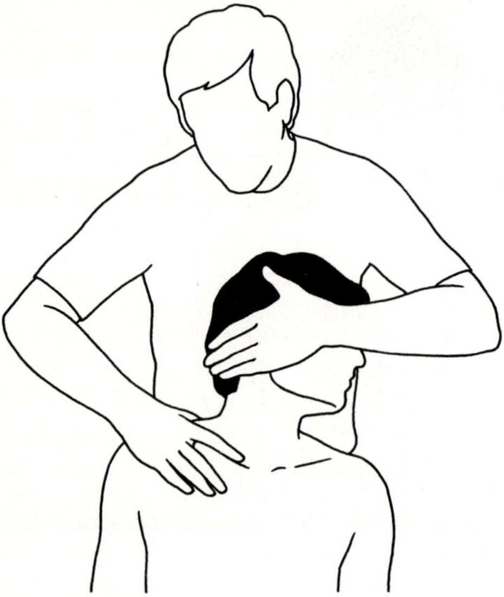

Fig. **24**

Execution Palpation of the scalene muscles in the region of their insertion, with the shoulder girdle immobilized. Passive extension and rotation of the cervical spine to the other side.

Interpretation A marked increase in the tone of the scalenes in the region of their insertions indicates shortening. Treatment on a trial basis will often confirm this difficult diagnosis.

4. Pectoralis major

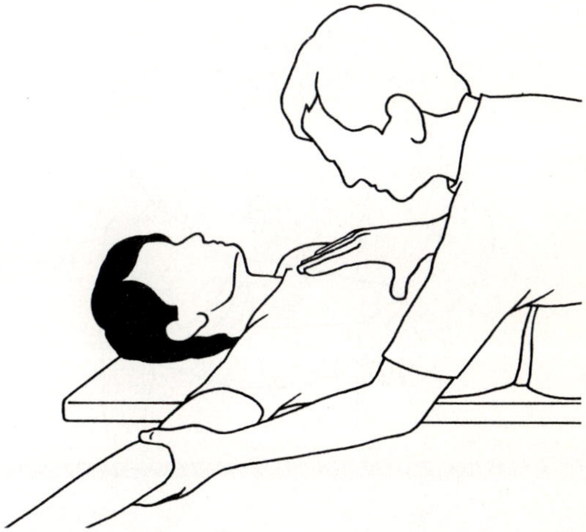

Fig. **25**

Execution	Abduction of the shoulder joint with the arm elevated at 120°. Avoid lordosation of the lumbar spine and rotation of the thoracic spine.
Interpretation	The arm should reach the level of the examination table. If it does not, and a soft stop is noted, the pectoralis major may be assumed to be shortened.

5. Extensores carpi et digitorum

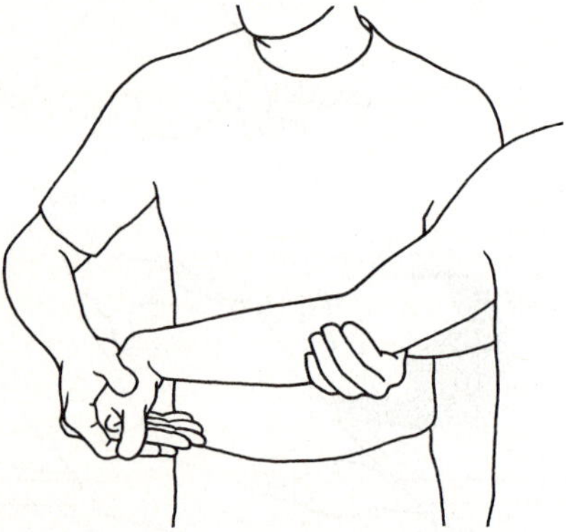

Fig. **26**

Execution The wrist and digits 2–5 are bent passively to maximum flexion and immobilized, with the elbow flexed at 45°. The elbow is extended passively while flexion of the wrist and fingers is maintained.

Interpretation Increasing pain in the proximal third of the forearm during extension indicates shortening. Pain in the distal third of the forearm and over the wrist may be due to tenosynovitis or synovitis of the wrist joint.

6. Erector spinae lumbalis

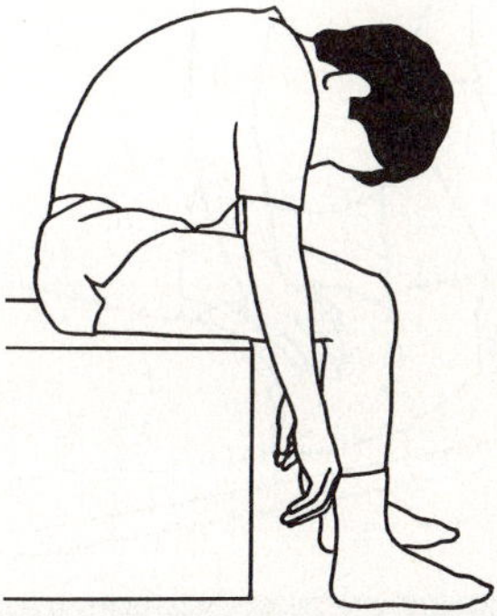

Fig. **27**

Execution Flexion of the lumbar and lower thoracic spine.

Interpretation Inflexibility of the lumbar spine and simultaneous bulging of the erector spinae in the lumbar region indicates shortening.

7. Psoas major

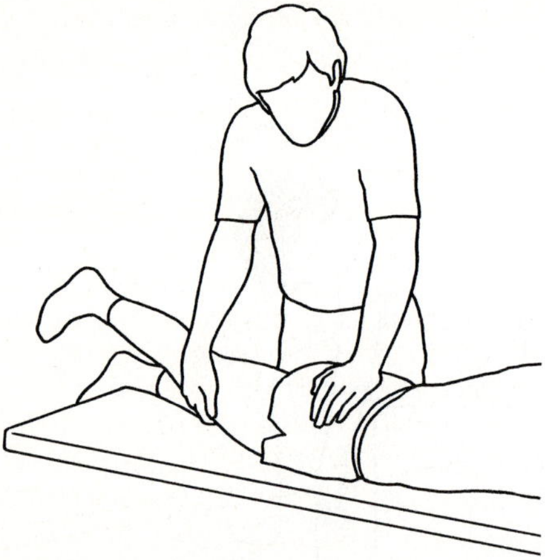

Fig. **28**

Execution	Extension of the hip joint with the pelvis firmly immobilized. Observe the transitional segments of the thoracolumbar spine.
Interpretation	Increasing lordosis cranial to the lumbar spine and lower thoracic spine indicates shortening of the psoas major.

8. Rectus femoris

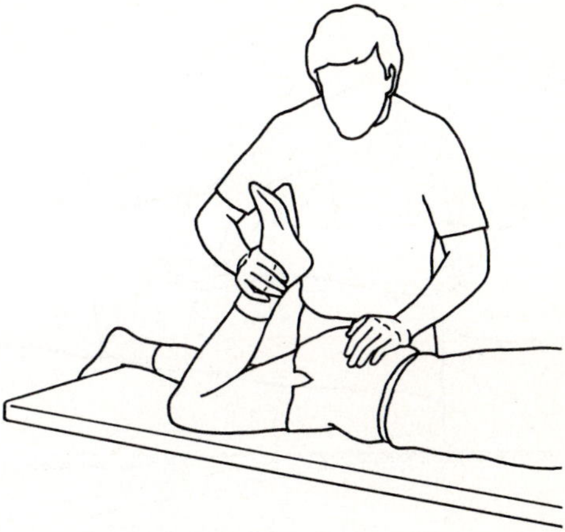

Fig. **29**

Execution Flexion of the knee joint. Yielding, palpating immobilization of the pelvis.

Interpretation Hip flexion occurring on flexion of the knee at an angle of less than 120° indicates a shortening of the rectus femoris. A hard reflex stop indicates nerve root compression cranial to nerve root L4.

9. Piriform muscle

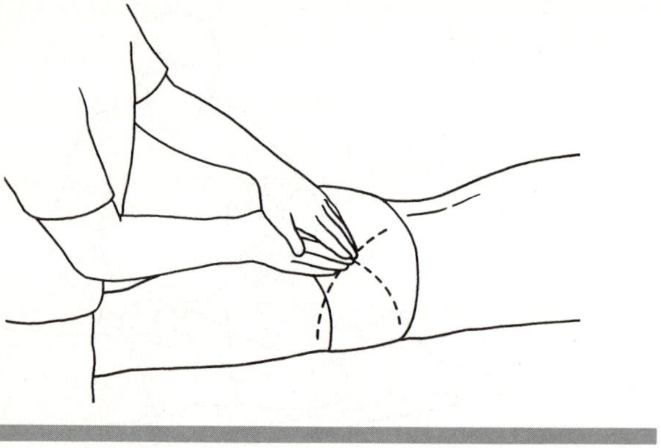

Fig. **30**

Execution Deep palpation of the gluteal muscle at the intersection of the lines joining the posterior superior iliac spine to the tip of the trochanter and the ischial tuberosity to the anterior iliac spine.

Interpretation A shortened piriform muscle can be readily palpated as a transverse cord. Palpation of the shortened muscle is painful. When not shortened, the piriform muscle can hardly be felt, or not at all.

10. Adductores femoris

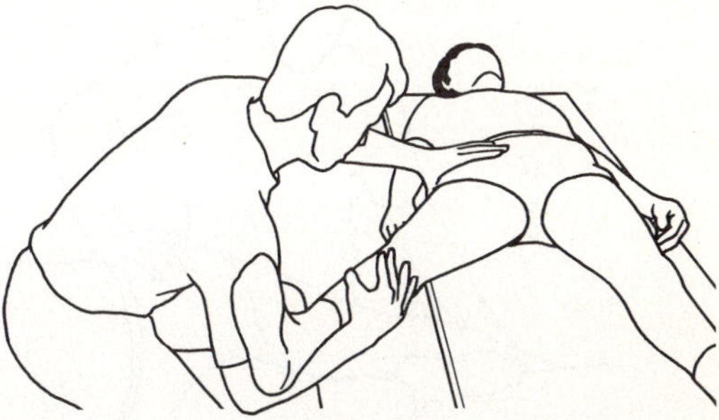

Fig. **31**

Execution Abduction of the hip joint. Immobilize and palpate the pelvis; avoid grasping the pes anserinus as a precaution against tactile stimulation of the adductor group.

Interpretation A soft stop at less than 45° abduction indicates shortening of the adductors.

11. Tensor fasciae latae

Fig. **32**

Execution Adduction of the hip joint. Good immobilization of the pelvis is essential.

Interpretation The middle of the tibia should reach at least to the level of the examination table. If it does not, and the skin over the iliotibial tract tightens, the tensor fasciae latae muscle may be assumed to be shortened.

12. Ischiocrural Muscles (biceps femoris, Semimembranosus, semitendinosus)

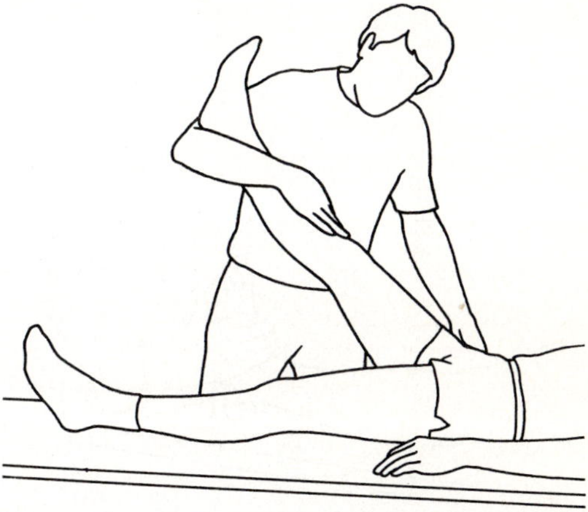

Fig. **33**

Execution Flexion of the hip joint. The arm gently supports the lower leg and controls rotation of the leg. Palpating immobilization of the pelvis with the proximal hand (important for evaluation of the stop).

Interpretation A soft stop at less than 80° hip flexion indicates a shortening of the ischiocrural muscles. A hard reflex stop is a sign of nerve root compression (Lasègue's sign). Differentiation from a hard reflex stop resulting from stimulation of the sacroiliac joint requires further investigation (neurological tests, examination of the sacroiliac joint, etc.). A hard stop due to arthrosis of the hip is not affected by simultaneous flexion of the knee, unlike the Lasègue's sign.

13. Triceps surae

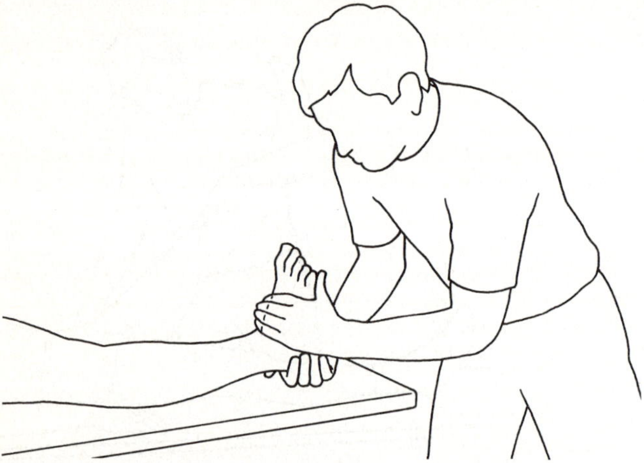

Fig. **34**

Execution	Dorsiflexion of the ankle joint with the knee extended. The foot is best grasped by the lateral border to avoid stimulation of the sole.
Interpretation	A soft stop at less than 20° dorsiflexion indicates a shortening of the triceps surae.

Treatment of Restricted Mobility

Physiological movement patterns require normal joint mobility and muscular balance.

Pathologically reduced joint mobility and muscular imbalance are treated by appropriate gymnastic exercises, unless, for example, provision of an endoprosthesis is indicated. It is absolutely essential to adapt the *intensity of exercise* to the *patient's clinical picture.*

Relief of pain is an important element of the *first phase* of the therapeutic program. Concomitant pain therapy can be carried out using physical methods (application of cold or heat, electrotherapy, hot baths, massage) or drug treatment (analgesics, nonsteroidal antirheumatics, local infiltrations).

The *second phase* consists of a course of physical therapy, often involving the application of a variety of techniques. Then the physician and physical therapist instruct the patient in an appropriate set of home exercises.

In the *third phase* the patient continues with the home exercises in order to maintain and possibly to improve on the therapeutic benefits of phases 1 and 2. For simple problems, the patient can be taught and started on a program of home exercises straight away.

Exercises employing techniques of *joint mobilization* and *muscle stretching* are used to improve mobility.

The patient's therapeutic program should be rounded out, in *phase 4,* with postural and movement training for work, sports, and leisure activities.

Treatment Plan for Restricted Mobility

Phase 1
- Treatment of pain, local or systemic
- Antirheumatics
- Analgesics, local infiltration
- Local physical therapy
- Medical gymnastics
- Manual therapy: manipulation and mobilization, traction

Phase 2
- Treatment of muscular imbalance
- Continuation of joint mobilization

Phase 3
- Home exercises

Phase 4
- Postural training, start or resumption of appropriate sport activities
- Promotion of fitness
- Prophylaxis of relapse by continued home exercise

Each phase should lead smoothly into the next. It is advisable to give about three treatments per week during phases 1 and 2.

Joint Mobilization Techniques

A distinction must be made between active and passive joint and vertebral mobilization.

Active mobilization techniques are suitable for a home exercise program.

The advantage of *active techniques of slow angular mobilization* is that they can be applied without irritating joint structures, as one has complete control over the force and direction of mobilization. To be of benefit, this type of mobilization must be preceded by optimum pretreatment of pain and combined with stretching excercises for any shortened tonic muscles.

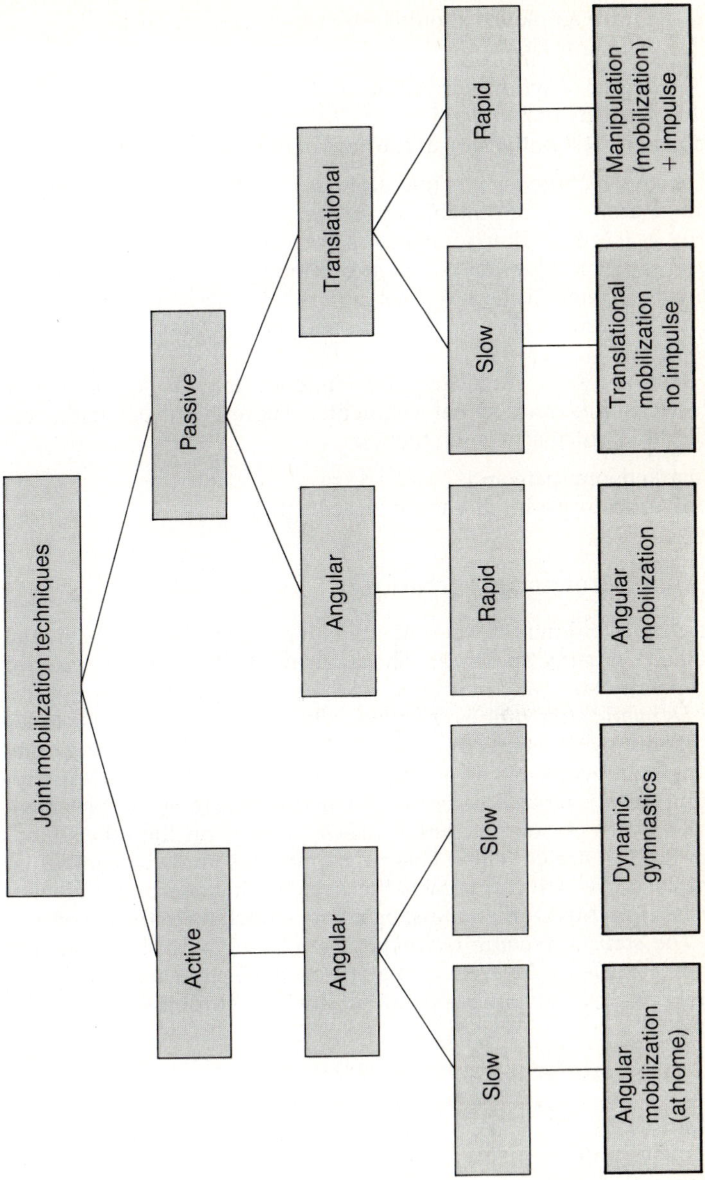

Fig. **35** Joint mobilization techniques

The force employed should be inversely proportional to

- the severity of arthrotic changes,
- the degree of synovial irritation,
- the intensity of pain, and
- the hardness of the stop at the limit of movement.

Active techniques of rapid angular mobilization, as in dynamic gymnastics, are not suitable for a therapeutic home exercise program, as it is difficult to regulate the force applied for mobilization.

Passive mobilization techniques belong for the most part to the domain of manual therapy; this is particularly true of *translational techniques* (references in W. Schneider et al.: Manual Medicine – Therapy).

Passive, slow mobilization techniques may be employed in selected joints and spinal segments. These home exercises are indicated particularly when there is

- marked joint pain and
- inflammatory joint disease in an acute phase.

Muscle Stretching Techniques

Basically two different types of stretching gymnastics can be distinguished: *dynamic* and *static.* The static methods are further divided into passive static and neuromuscular stretching techniques.

Dynamic stretching is the type most commonly practiced today in sports gymnastics. By means of rebounding, bouncing, and swinging movements, an attempt is made to stretch the corresponding muscles as far as possible and to increase the range of movement of the joints. A sudden, jerky muscle stretch, on the other hand, activates the stretch reflex. This reflex produces immediate muscular contraction, which counteracts the stretch. This neurophysiological mechanism thus works against optimum stretching of the muscles.

The static stretching technique ("stretching") can be applied to obtain complete relaxation of the muscle to be stretched either with purely passive movements or by employing neuromuscular mechanisms.

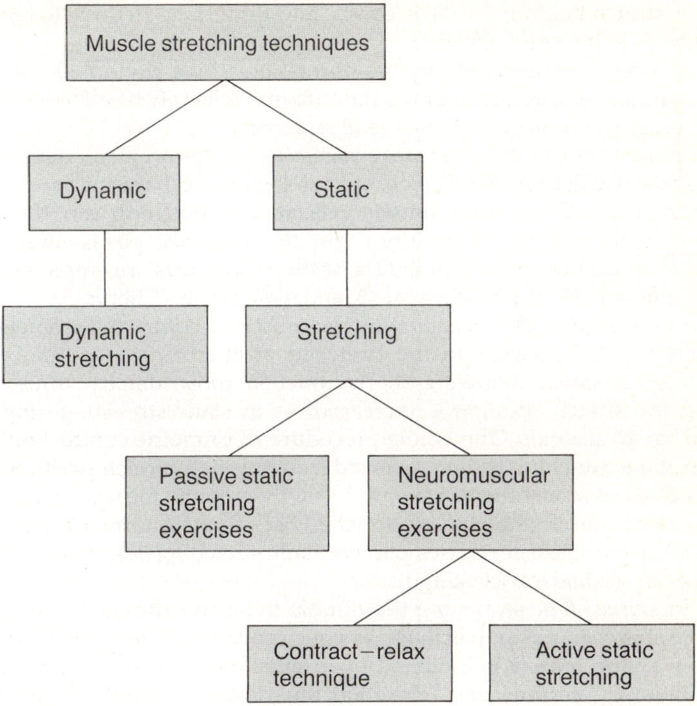

Fig. **36** Muscle stretching techniques

In purely *passive static stretching* – the most common form of stretching – the muscle is brought into the extended position and then stretched further by a small change in position. This change in position can be effected by gravity, the subject's own muscular power, a partner, or apparatus.

By changing the stretch position, a gradual increase in resistance is obtained. The subject stops at the position in which the stretch sensation is still pleasant. Mild tension of the muscle is permitted, but the stretch should not be painful. Recognizing the correct stretch sensation requires some degree of practice. Stretching is accordingly something that has to be learned.

One objective (not always attainable) of this slow stretching is to avoid activating the stretch reflex and thus to perform the stretch on a relaxed muscle without the occurrence of disturbing reflex muscle contractions.

A stretch held for 15–30 seconds, and performed in three to ten stages, is sufficiently effective.

During the stretch phase, the normal breathing rhythm should be maintained. Stretching of the musculature will only be effective if sufficient attention is paid to general relaxation.

In *neuromuscular stretching exercises* (proprioceptive neuromuscular facilitation, PNF), neurophysiological mechanisms are deliberately used to obtain muscle relaxation. Stretching can then occur under optimum conditions. For this purpose, postisometric inhibition and reciprocal inhibition of the antagonists are employed (references in W. Schneider et al.: Manual Medicine – Therapy).

In the *contract-relax technique* the muscle is contracted isometrically for 3–7 seconds in the stretched position. In the phase of muscle relaxation following the contraction (postisometric inhibition) the stretch position is increased, as in static stretching, and held for 10 seconds. The whole procedure of isometric contraction/relaxation/stretching is then repeated from this new stretch position. This form of stretching can be used as a therapeutic measure when shortened muscles need to be stretched back to their normal length. For daily gymnastic exercises passive static stretching is sufficient to maintain normal muscle length.

In *active static stretching* the muscle to be stretched is brought into a stretched position actively by contraction of its antagonists. In this way the muscle is inhibited by a reflex mechanism (reciprocal inhibition). The muscular relaxation thus obtained facilitates optimum stretching. The duration of the stretch phase is 10–20 seconds.

To obtain and maintain good muscular balance, stretch exercises for the tonic muscles and strengthening exercises for the phasic muscles should be built into the gymnastics routine. Exercises to improve strength will be presented subsequently in Part 2 of this program ("Strength"). In some cases, long-term concomitant treatment of pain may be necessary.

Tips for Teaching Home Exercises

Which exercises should be taught?

Examination of your patient has revealed limited mobility of a joint or spinal segment. You will choose the most appropriate mobilization or stretching exercise according to the type of stop at the limit of movement. In the case of a soft stop, stretching of the shortened muscle is recommended; in the case of a hard stop, mobilization of the joint should be implemented.

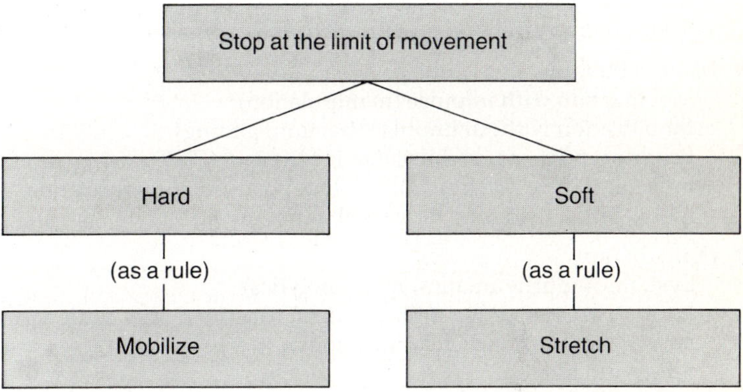

Fig. **37**

The mobilization exercises are done over the whole range of motion. Stretching exercises go from the position of maximum stretch in the direction of the limitation of movement.

How many exercises should you teach your patient?

We suggest that you build up a home exercise program progressively. Two to four exercises can be taught at each session. Further exercises are added as necessary, but only when you are sure that the previous exercises have been mastered.

What needs to be done if pain occurs during or after the exercises?

Mobilization exercises, if performed very energetically, can produce painful reactions. Stretch exercises, on the other hand, should not cause pain afterwards.

Apart from modifying exercises, supplementary measures may be necessary:

- Modification of exercises
  - More cautious application of force, especially toward the end of the movement
  - Choice of less painful starting positions
  - A change in the direction of mobilization and stretching
- Physical therapy
  - Cryotherapy and thermotherapy
  - Hydrotherapy
  - Electrotherapy
  - Reflex therapy (massage, acupuncture, acupressure)
- Manual therapy
  - Mobilization with impulse (manipulation)
  - Mobilization without impulse (traction, gliding)
  - Proprioceptive neuromuscular facilitation (PNF): using direct muscular force – stretching using postisometric relaxation – mobilization using the reciprocal inhibitory effect
- Drug therapy
  - Systemic (antirheumatics, myotonolytics)
  - Local (percutaneous, infiltration of muscle points with local anesthetics, intra-articular injections)
  - Spray and stretch

When is it advisable to have a physical therapist instruct the patient?

In cases where there is a marked loss of mobility, especially when combined with severe pain, an individualized introduction to therapy and instruction by a physical therapist can be valuable.

If you yourself are unsure of the procedure in particular exercises, have a therapist support you with instruction and surveillance (be sure to set out your therapeutic objectives and wishes clearly).

It is also useful to call in a physical therapist when reduced mobility is accompanied by muscular weakness and/or abnormal movement and posture patterns.

What advice can you give your patient for the mobilization exercises?

- Assume starting positions as shown.
- Always cover the full range of movement slowly and smoothly.
- Apply force gradually.

- Repeat each movement five to ten times.
- Do not be put off by mild joint pain.

What advice can you give your patient for the stretching exercises?

- Assume starting positions as illustrated.
- Change your position slowly in the direction of the arrow to increase the stretch.
- Avoid sudden movements (no bouncing). A slight sensation of pulling in the muscles being stretched is normal.
- Hold each stretch position for 15–30 seconds.
- Repeat exercises two or three times, breathing slowly and evenly throughout.

What advice do we have for you?

Familiarize yourself with the exercises by including them now and again in your own wellness or fitness program. And if you can get someone to join you (be it spouse, teammate, colleague, or another physical therapist), so much the better.
Good luck!

Systematic Exercise Program

1. Cervical Spine

Technique Mobilization

Execution ↗↙ Draw in your chin (nodding movement).

Note – This is a small movement, primarily of the upper cervical spine.

2. Cervical Spine

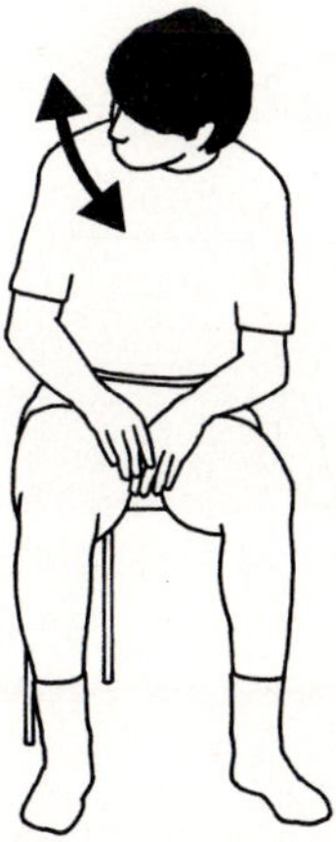

Technique Mobilization

Execution Flex your neck as far as possible and turn your head to the side.

Note – The turning movement occurs chiefly in the upper cervical spine.

3. Cervical Spine

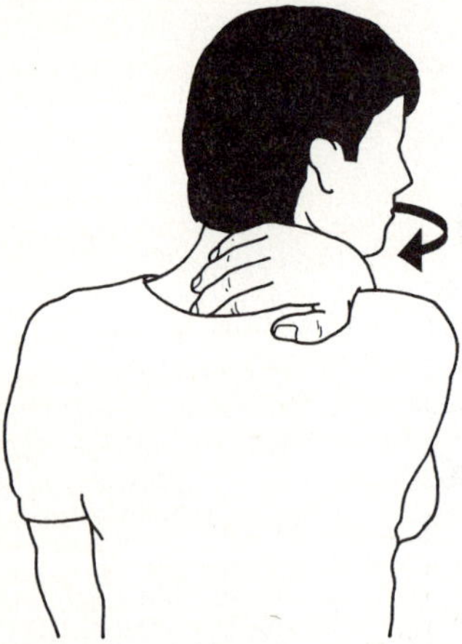

Technique Mobilization

Execution ↷ Immobilize the lower segments of the cervical spine with one hand. Turn your head toward the same side.

Note – The movement occurs chiefly in the middle and upper cervical spine.

4. Cervical Spine

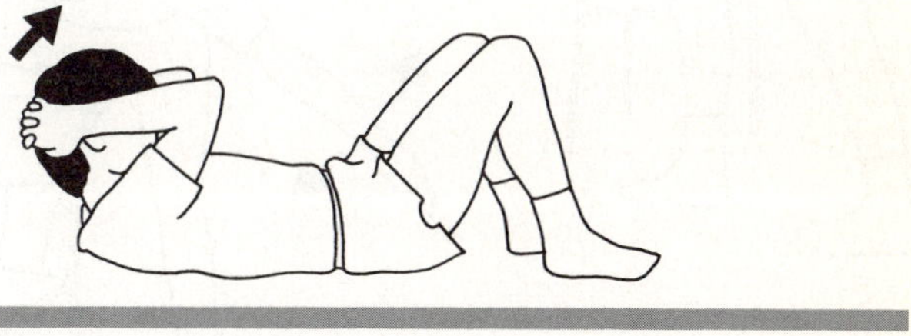

Technique Passive static stretch

Execution ↗ Slowly pull your head forward with both hands and breath out (stretch for 10 seconds).

Notes
- The head should be pulled in the direction of its long axis ("off the neck") at the same time as it is pulled forward.
- This exercise is often good for a neckache or headache.
- Stop doing this exercise at the first sign of dizziness.
- Because of the stress it puts on disks and ligaments, this exercise should not be tried for several weeks after sprain of the cervical spine.

5. Shoulder Joint

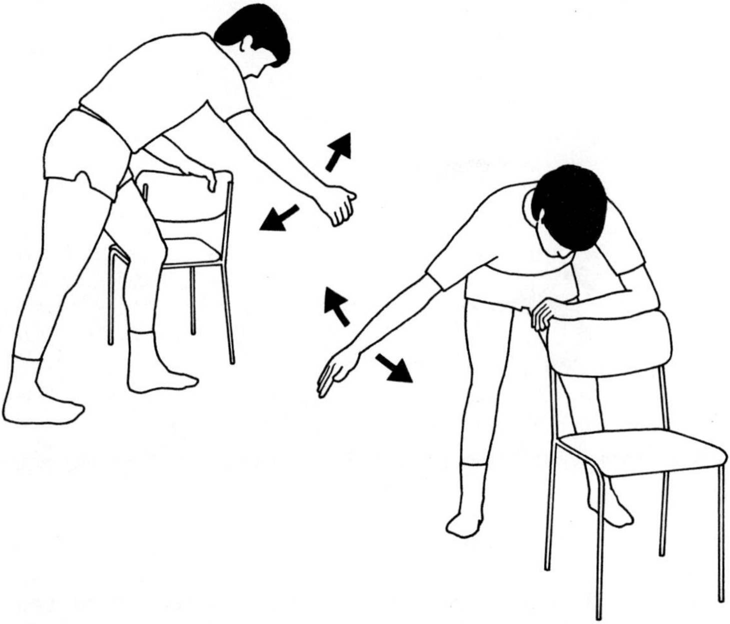

Technique	Mobilization
Execution	Swing your arm back and forth.
	Swing your arm out to the side.
Note	– The movement can be performed in several forward-bending positions.

6. Shoulder Joint

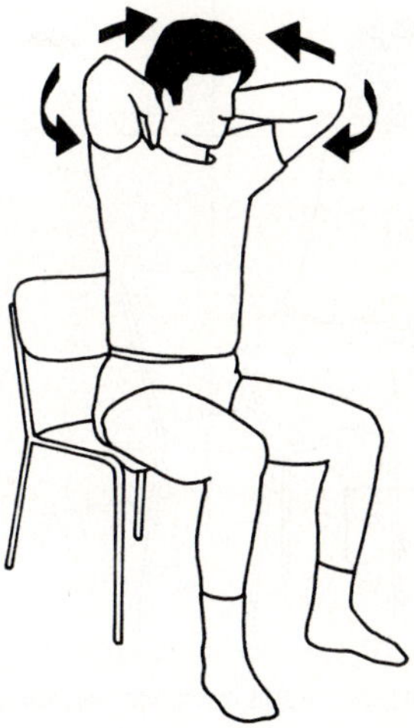

Technique	Mobilization
Execution	Move your elbows back and forth.
Note	– Do not press your head forward.

7. Shoulder Joint

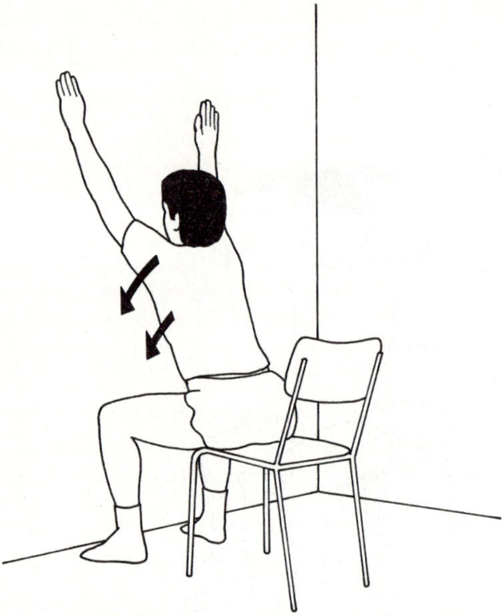

Technique	Mobilization
Execution ✓	Press the upper part of your body forward and down.
Note	– Vary the distance between your hands.

8. Shoulder Joint

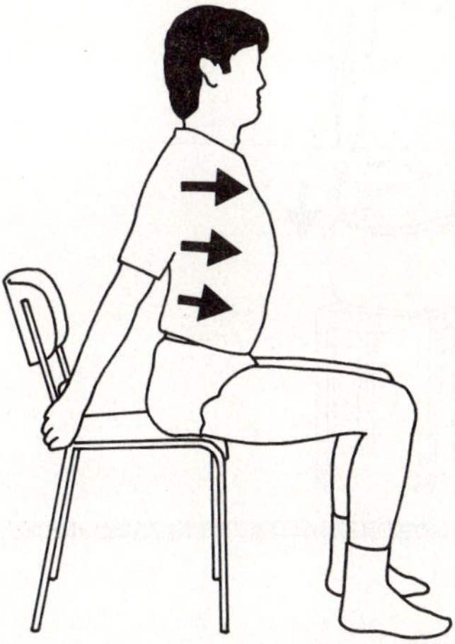

Technique	Mobilization
Execution	Tip your pelvis forward and stretch your back.
Note	– Mobilize as you exhale.

9. Shoulder Joint

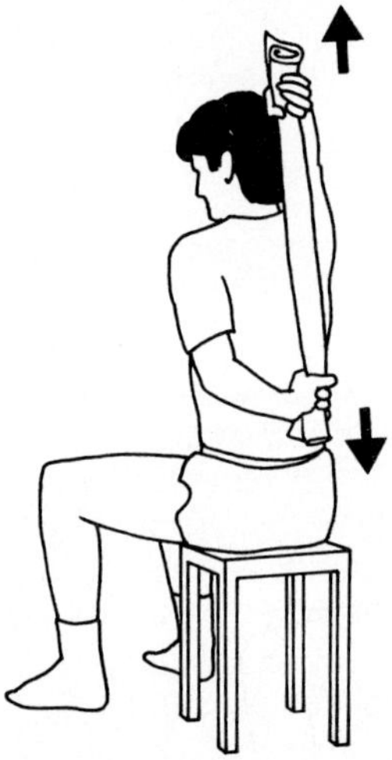

Technique	Mobilization
Execution ↕	Pull the towel up and down.
Note	– Keep your back straight and press your elbows back.

10. Shoulder Joint

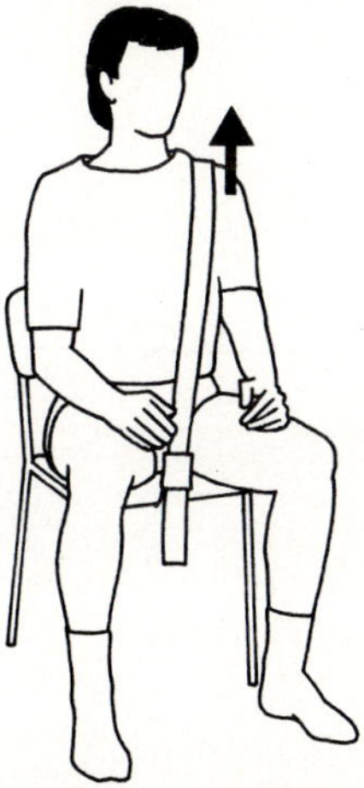

Technique	Mobilization
Execution ↑	Pull your shoulder up.
Notes	– The strap is placed over the outer end of the clavicle. – Apply force gradually.

11. Shoulder and Neck Muscles

Technique Passive static stretch

Execution ↘ Tilt your head to one side.

↘ Pull down the opposite arm and breathe out.

Note – Keep your upper body straight.

12. Shoulder and Neck Muscles

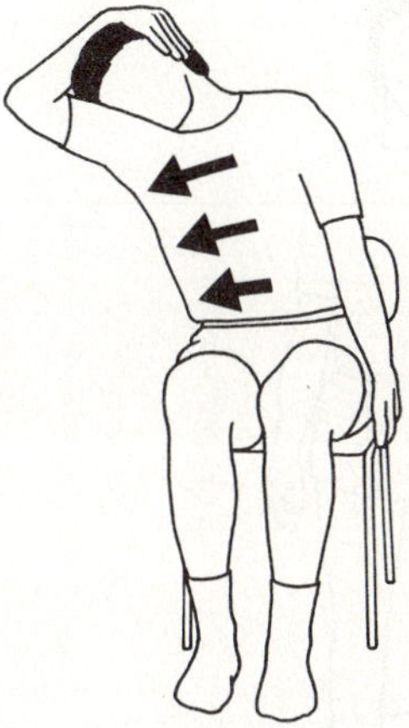

Technique Passive static stretch

Execution Tilt your head to one side and immobilize it with your hand. Hold on to the seat of the chair with the opposite hand. Slowly bend your trunk to the side and breathe out.

Note
- When the exercise is done correctly, the shoulder and neck muscles stretch entirely as a result of lateral bending of the trunk. Do not pull on your head.

13. *Pectoral Muscles*

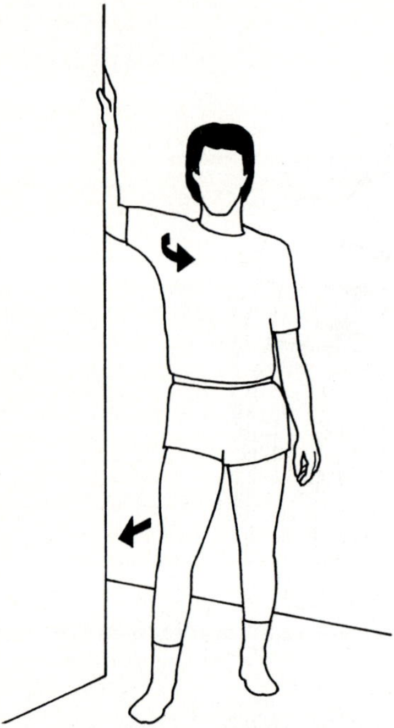

Technique Passive static stretch

Execution Step forward with your inside leg. Shift the shoulder on the same side forward.

Note – Different parts of the pectoral muscles can be stretched by altering the elevation of the arm.

14. Posterior Upper Arm Muscles

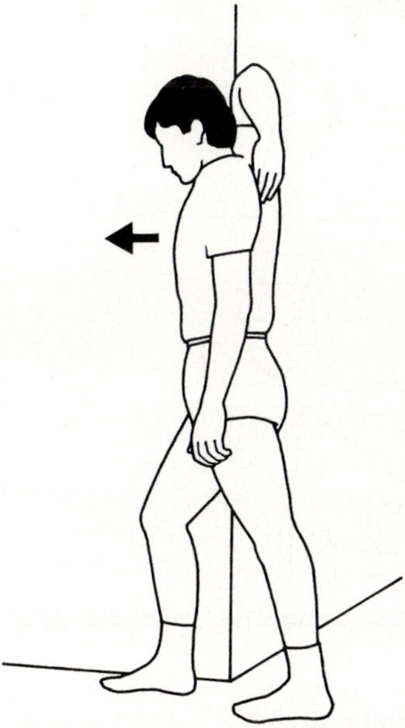

Technique	Passive static stretch
Execution ←	Move your trunk forward with your arm raised and bent back.
Note	– An important stretching exercise for restoring full extension of the arm.

15. External Forearm Muscles

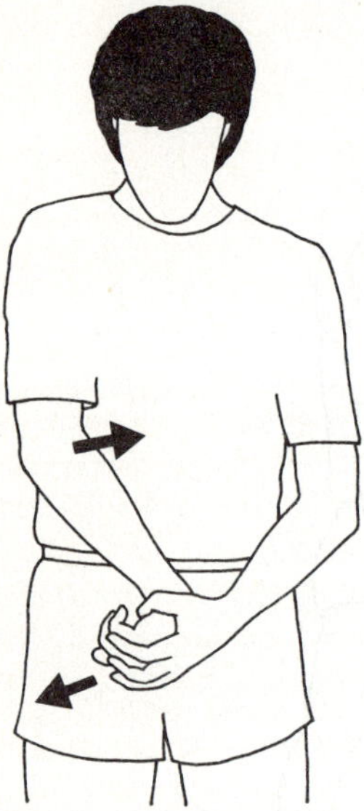

Technique	Passive static stretch
Execution	Immobilize your hand with your wrist and fingers flexed.
⇄	Extend your elbow.
Note	– The more firmly the wrist and finger joints can be immobilized in the flexed position, the more intensive the muscle stretch will be.

16. Hand

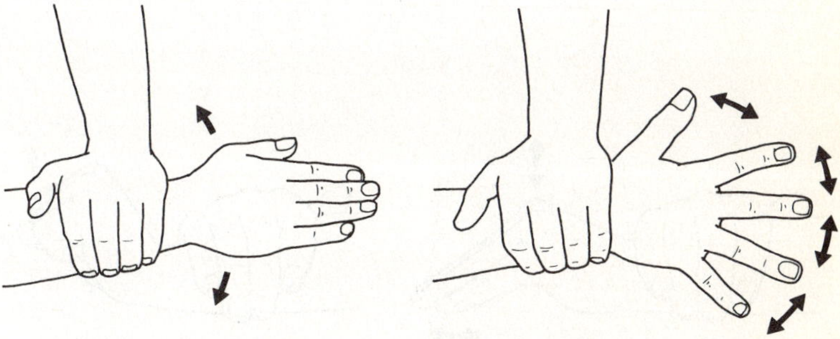

Technique	Mobilization
Execution	↖↘ Flex your wrist sideways.
	↕ Open and close your fingers.
Note	– Keep your forearm from sliding by pressing down with your other hand.

17. Hand

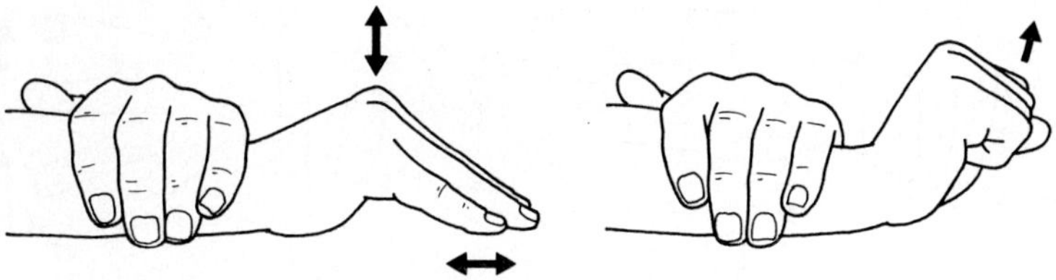

Technique	Mobilization
Execution ↕ ↔	Extend your fingers on a flat surface and draw them back without bending them.
↑	Clench and open your fist.
Note	– Keep your forearm from sliding by pressing down with your other hand.

18. Thumb Joint

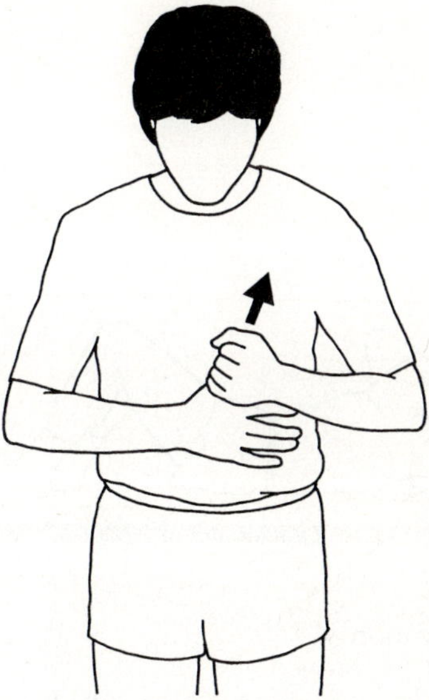

Technique	Mobilization
Execution	➚ Rest your hand against your body. Grasp your thumb and pull it along its axis.
Note	– Apply gradually increasing traction for five seconds.

19. Thoracic Spine

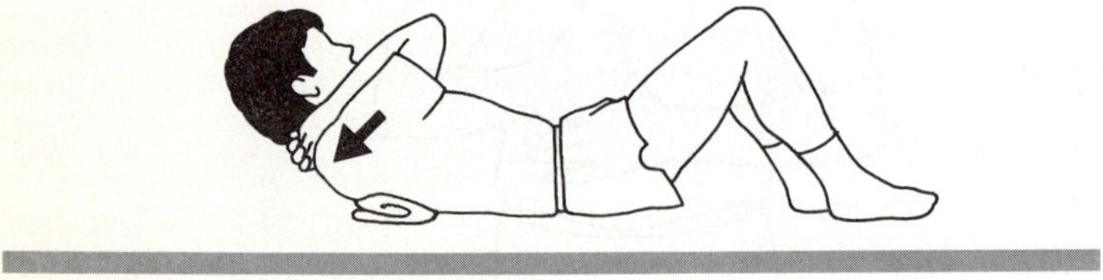

Technique Mobilization

Execution ↙ Roll your spine slowly over a rolled-up towel.

Note – The position of the towel determines at what level the spine is mobilized.

20. Thoracic and Lumbar Spine

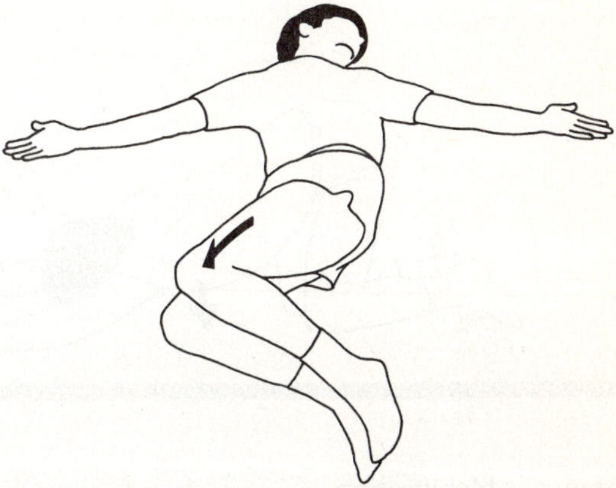

Technique Mobilization

Execution ↙ Lie on your back with knees bent and lower your knees to one side.

Turn your head to the opposite side.

Note – The shoulder girdle must remain in contact with the mat.

21. *Thoracic and Lumbar Spine*

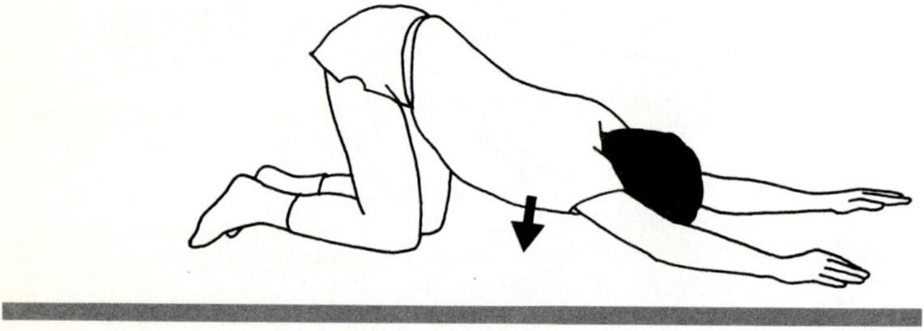

Technique Mobilization

Execution ↓ Press your upper body downward.

Notes
- Do not let your head hang down.
- Mobilize as you exhale.

22. Thoracicx and Lumbar Spine

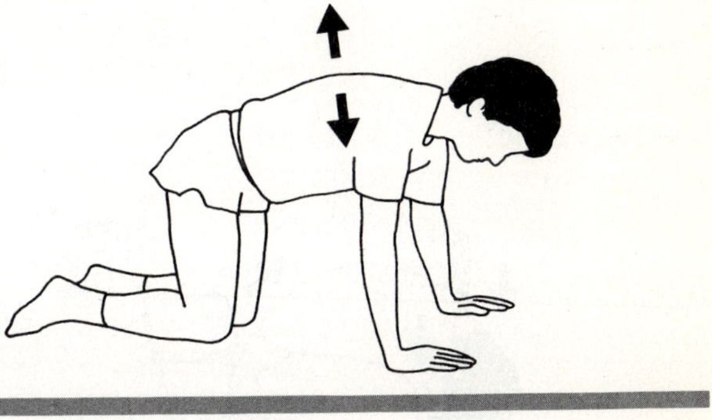

Technique Mobilization

Execution ↕ Alternately arch and stretch your back.

Note – Do the exercise in time with your breathing.

23. Thoracic and Lumbar Spine

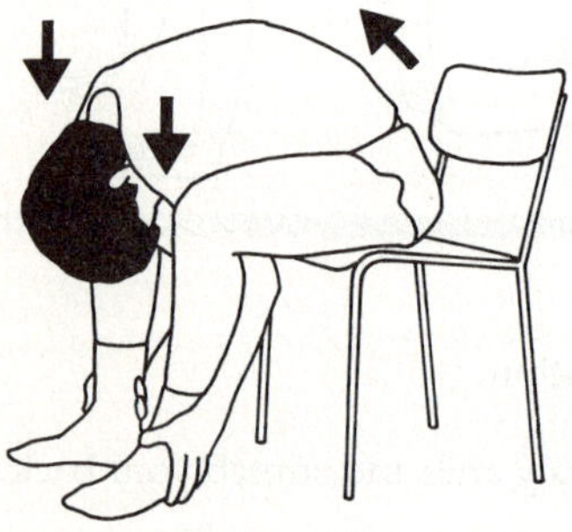

Technique	Passive static stretch
Execution	Curve your back by pulling down with your arms.
Note	– If the exercise is done correctly, you will feel the muscles stretching in your lower back. The main objective is to stretch the extensor muscles of the lumbar region.

24. Lateral Muscles of the Trunk

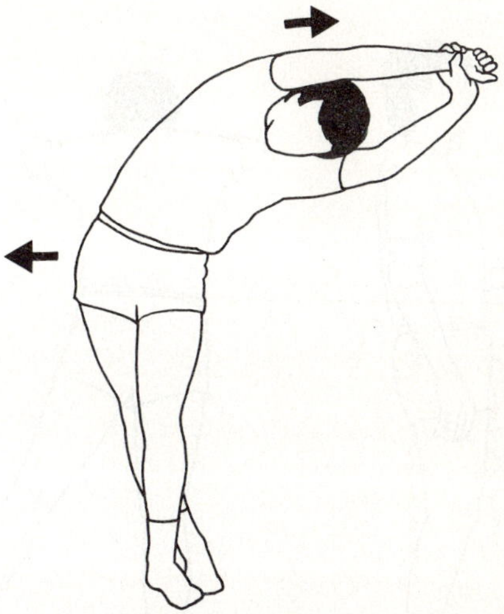

Technique	Passive static stretch
Execution	← Push your hip out sideways. → Bend your trunk to the other side and pull.
Notes	– Put your left foot forward before bending to the left, and vice versa. – The two movements should be in the same plane.

25. Hip Joint

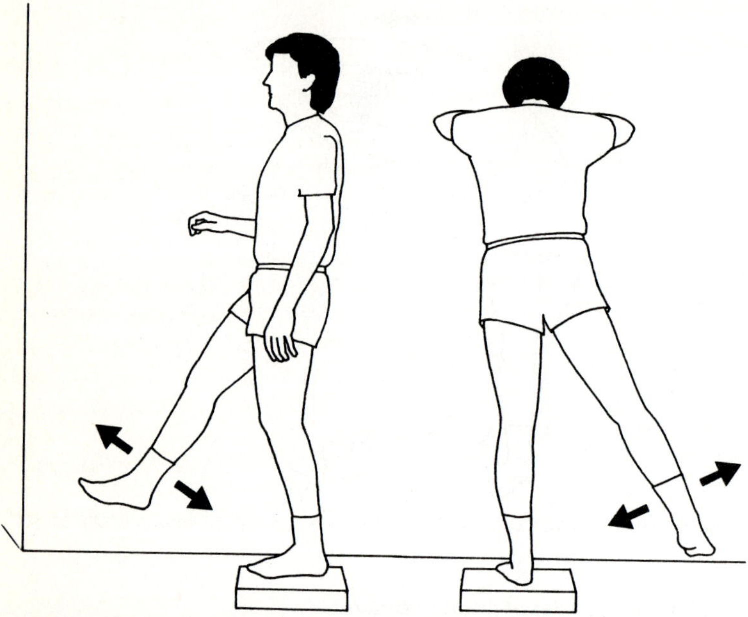

Technique Mobilization

Execution ↖↘ Swing your legs back and forth.
↙↗ Now swing them out to the side.

Note – Your upper body should not move during the exercise.

26. Hip Joint

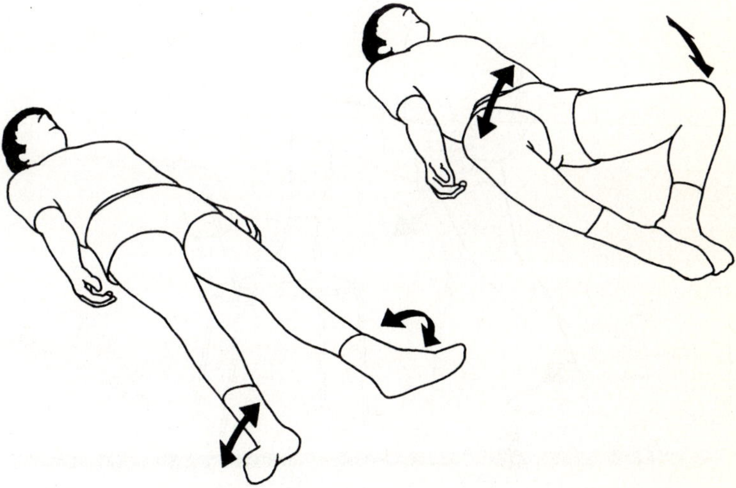

Technique	Mobilization
Execution ↶	Turn your legs outwards and inwards.
↕	With knees bent, open and close your legs.
Notes	– In the first half of the exercise your legs should be slightly apart.
	– In the second, your feet should be touching.

27. Anterior Hip Muscles

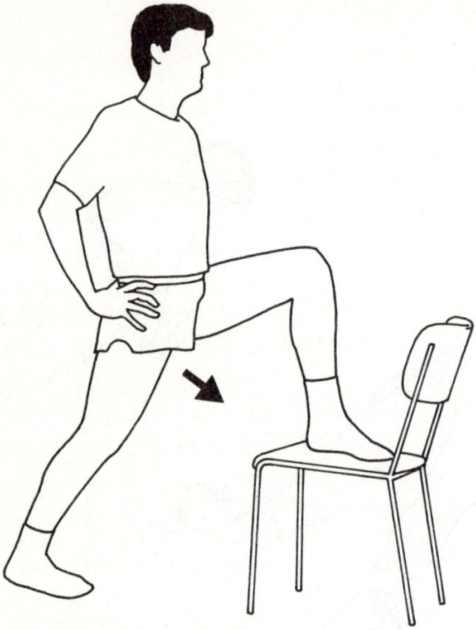

Technique	Passive static stretch
Execution ➘	Push your hips forward.
Note	– For specific stretching of the hip flexors, do not turn your hips outwards. Your feet should point forward.

28. Internal Hip Muscles

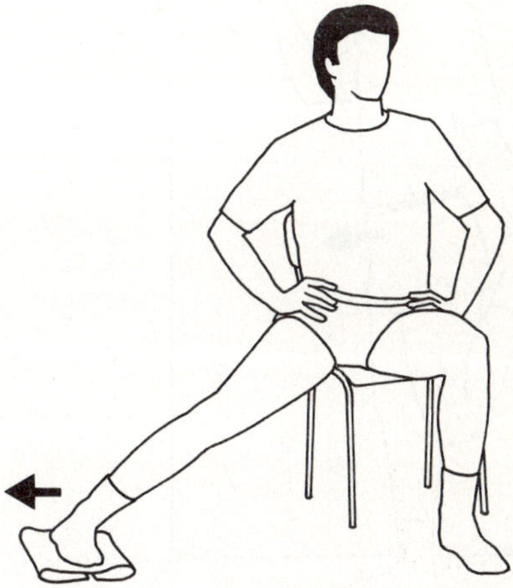

Technique Passive static stretch

Execution ← Push your heel out to the side.

Notes
- The toes should point forward.
- The knee should be straight.
- If mobility is severely limited, the seat must be higher.

29. Anterior Thigh Muscles

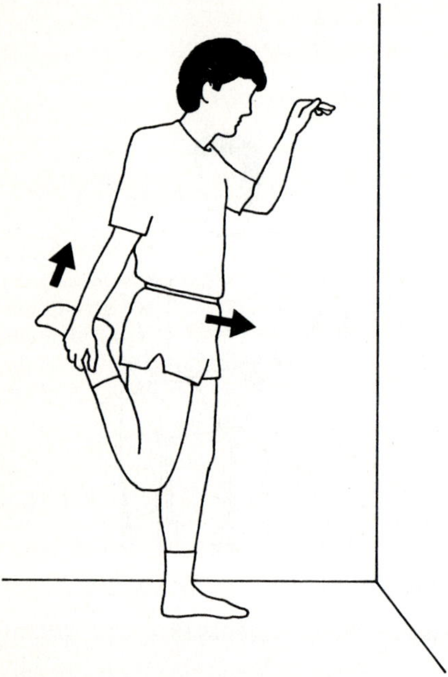

Technique		Passive static stretch
Execution	↑	Pull your foot towards the buttock.
	→	Push your pelvis forward.
Notes		– By pulling up on the forefoot, you stretch the anterior muscles of the lower leg at the same time.
		– Do not try to compensate by hollowing your back.

30. Posterior Thigh Muscles

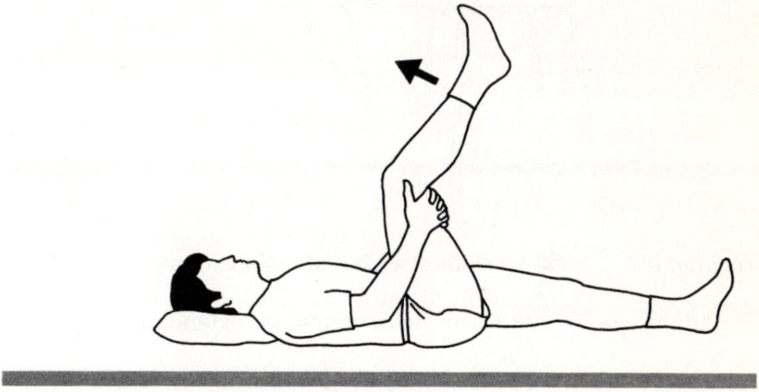

Technique Active static stretch

Execution ↖ Immobilize your thigh with the hip bent and actively straighten the knee.

Note – The hip remains flexed during the stretch.

31. Posterior Hip Muscles/Sacroiliac Joint

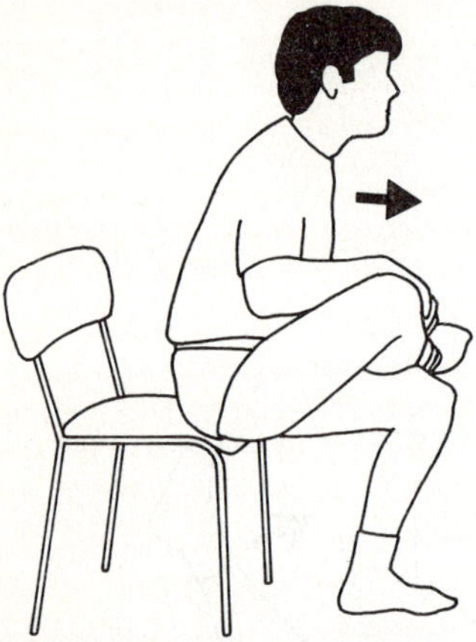

Technique Passive static stretch and mobilization

Execution ➔ Bend forward with your spine extended.

Note – Stop stretching if you notice any abnormal sensation in your legs.

32. Anterior Thigh Muscles

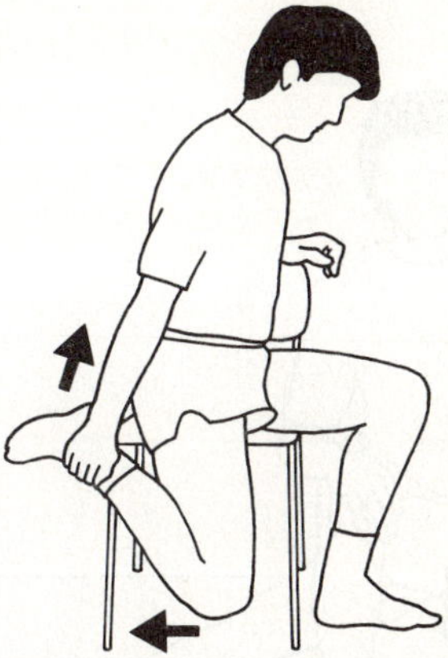

Technique	Passive static stretch
Execution ↑	Pull your heel towards the buttock.
←	Slowly move the knee back.
Notes	– Your head should be bent slightly forward.
	– If shortening is severe, use a towel to hold onto the foot.

33. Lateral Hip Muscles

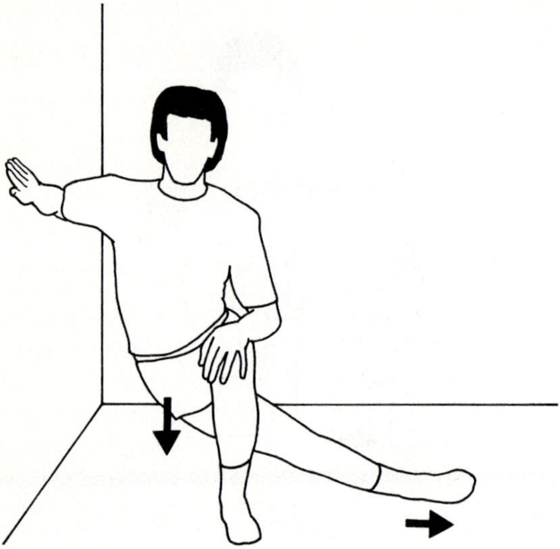

Technique	Passive static stretch
Execution ↓→	Push your pelvis downward.
Note	– The upper part of the body must be well supported.

34. Posterior Thigh Muscles

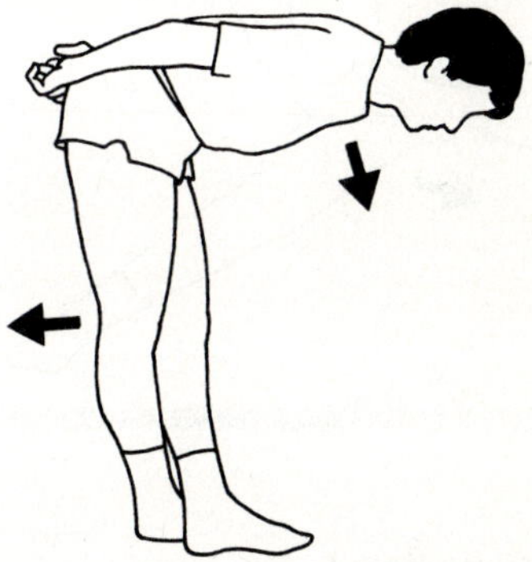

Technique	Passive static stretch
Execution ←	Stretch your knee.
↘	Bend forward.
Notes	– The back should be kept as straight as possible. – A difference in tension between the two sides will indicate which side needs to be stretched more intensively.

35. Internal Hip Muscles

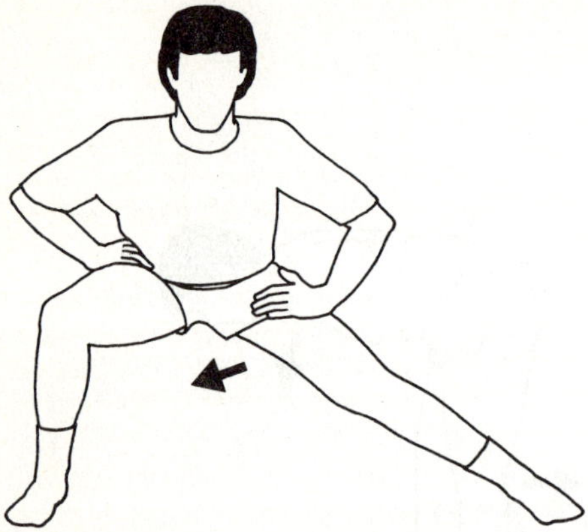

Technique	Passive static stretch
Execution ↙	Push your pelvis down and to the inside.
Note	– An easier version: use both hands to support the weight of your upper body on your bent knee or the seat of a chair.

36. Anterior Hip Muscles

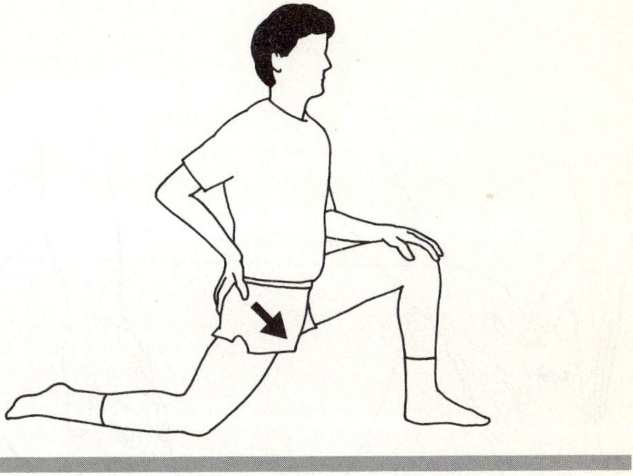

Technique	Passive static stretch
Execution ➘	With one hand, push your pelvis forward and down.
	Press your hip forward and down.
Notes	– Keep your back straight.
	– Your feet should point straight forward and back.

37. Knee

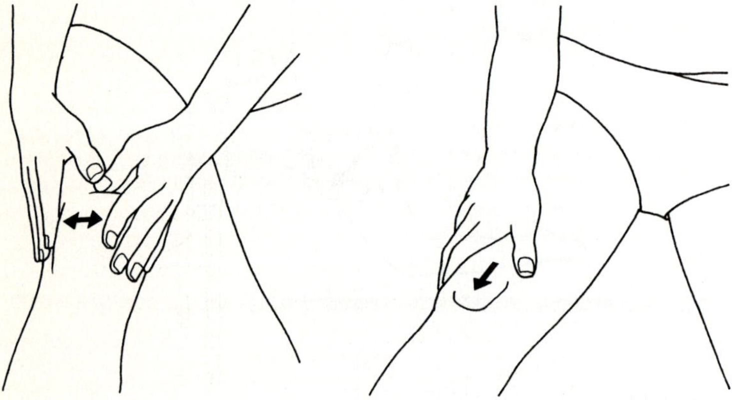

Technique		Mobilization
Execution	↙	Push your kneecap down toward the foot.
	↔	Now push it sideways in both directions.
Note		– Do not press down on the kneecap.

38. Posterior Lower Leg and Thigh Muscles

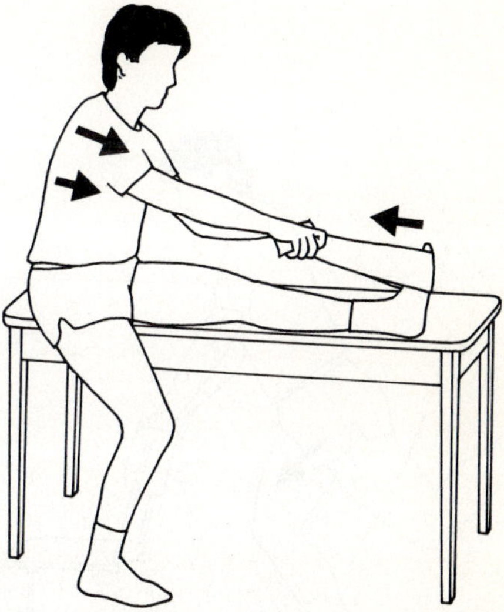

Technique Passive static stretch

Execution ← Pull back on your forefoot with a towel.

→ Lean forward.

Note – The knee is extended.

39. Anterior Thigh Muscles

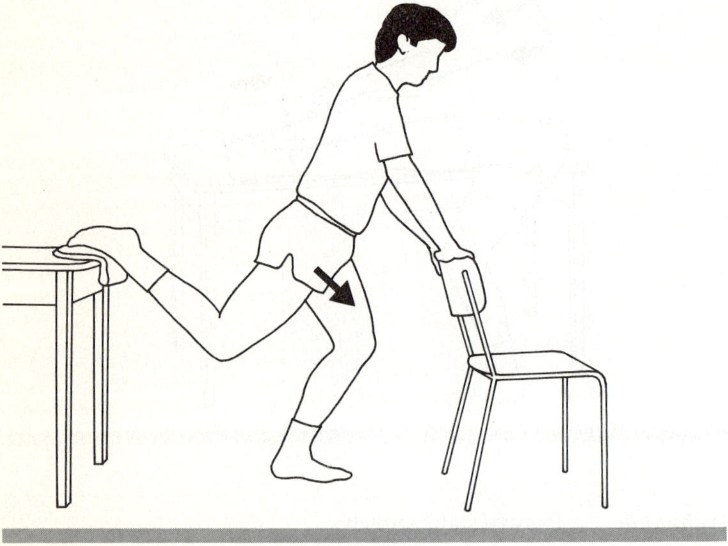

Technique	Passive static stretch
Execution ➘	Bend the knee of the supporting leg and push your hip forward and down.
Note	– In this exercise the rectus femoris (the straight muscle of the thigh) is stretched in the region of the hip and very little in the region of the knee. Since very little load is placed on the knee joint, the exercise can be done even when the knee is painful.

40. Posterior Lower Leg Muscles

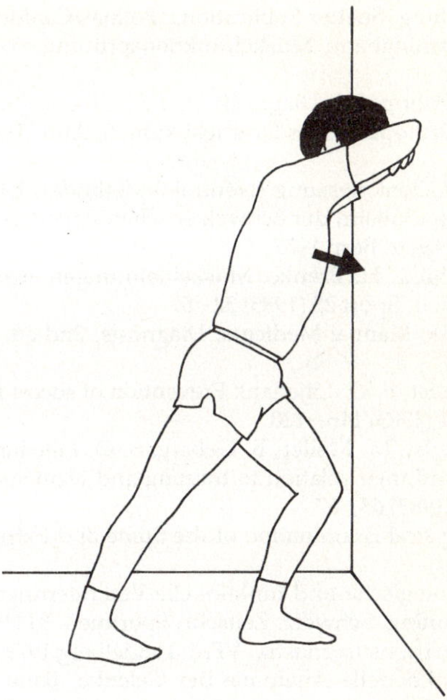

Technique	Passive static stretch
Execution	Press your heel to the ground.
	➘ Gradually lean forward.
Note	– The deep muscle of the calf (soleus muscle) can be stretched simply by bending the front knee as well.

References

Anderson, B.: Stretching. Shelter Publications, Bolinas/California 1980

Daniels, L., C. Worthingham: Muskelfunktionsprüfung, 5. Aufl. Fischer, Stuttgart 1985

Debrunner, H. U.: Orthopaedic Diagnosis, 2nd ed. Thieme, Stuttgart 1982

Debrunner, H. U.: Orthopädisches Diagnostikum, 5. Aufl. Thieme, Stuttgart 1987

Debrunner, H. U.: Gelenkmessung (Neutral-0-Methode), Längenmessung, Umfangmessung. Bulletin der Schweizerischen Arbeitsgemeinschaft für Osteosynthesefragen, Bern 1977

Dietrich, L., F. Berthold, H. Brenke: Muskeldehnungen aus sportmethodischer Sicht. Med. u. Sport 25 (1985) 52–57

Dvořák, J., V. Dvořák: Manual Medicine, Diagnosis, 2nd ed. Thieme, Stuttgart 1990

Ekstrand, J., J. Gillquist, S. O. Liljedahl: Prevention of soccer injuries, Amer. J. Sports Med. 11 (1983) 116–120

Ekstrand, J., J. Gillquist, M. Möller, B. Oeberg, S. O. Liljedahl: Incidence of soccer injuries and their relation to training and team success. Amer. J. Sports Med. 11 (1983) 63–67

Hoppenfield, S.: Physical Examination of the Spine and Extremities. Appleton, New York

Howald, H.: Morphologische und funktionelle Veränderungen der Muskelfasern durch Training. Schweiz. Zeitschr. Sportmed. 31 (1984) 5–14

Janda, V.: Muskelfunktionsdiagnostik. VFM, Heidelberg 1979

Kapandji, I. A.: Funktionelle Anatomie der Gelenke, Band 1, 2, 3. Enke, Stuttgart 1984/1985

Kelly, W., E. Harris, S. Ruddy, C. Sledge: Textbook of Rheumatology, 2. Aufl. Saunders, Philadelphia 1985

Knebel, K. P.: Funktionsgymnastik. Rowohlt, Reinbek 1985

Mellerowicz, H., W. Meller: Training, 4. Aufl. Springer, Berlin 1980

Moore, M. A., R. S. Hutton: Electromyographic investigation of muscle stretching techniques, Med. Sci. Sports 12 (1980) 322–329

Schmid, H., H. Spring: Muscular imbalance in skiers, Man. med. 21 (1983) 63–66

Schmidt, R. F.: Grundriß der Neurophysiologie, 4. Aufl. Springer, Berlin 1979

Schneider, W.: Stretching and Isometrics, Roche, Basel 1984

Schneider W., J. Dvořák, V. Dvořák, T. Tritschler: Manual Medicine: Therapy. Thieme, Stuttgart 1988

Sölveborn, S. A.: Das Buch vom Stretching. Mosaik, München 1983

Spring, H.: Muskelfunktionsdiagnostik nach Janda, Ergebnisse einer Untersuchung an Skirennfahrern. Schweiz. Zeitschr. Sportmed. 29 (1981) 143–146

Spring, H.: Was bringt das Stretching? Schweiz. Zeitschr. Sportmed. 33 (1985) 21–24

Spring, H., W. Schneider, T. Tritschler: Erkennung und Verhütung muskulärer Dysbalancen, Therapeutische Umschau 44 (1987) 868–877

Spring, H., U. Illi, H. R. Kunz, K. Röthlin, W. Schneider, T. Tritschler: Stretching and Strengthening Exercises. Thieme, Stuttgart 1991

Stegemann, J.: Leistungsphysiologie, 3. Aufl. Thieme, Stuttgart 1984

Tittel, K.: Beschreibende und funktionelle Anatomie des Menschen, 9. Aufl. Fischer, Stuttgart 1981

Uram, P.: The Complete Stretching Book. Anderson World, Mountain View/ California 1980

Wallin, D., B. Ekblom, R. Grahn, T. Nordenborg: Improvement of muscle flexibility, a comparison between two techniques. Amer. J. Sports Med. 13 (1985) 263–268

Weber, J., F. Berthold, H. Brenke, L. Dietrich: Die Bedeutung muskulärer Dysbalancen für die Störung der arthromuskulären Beziehungen. Med. u. Sport 25 (1985) 149–151

Weineck, J.: Optimales Training. Perimed, Erlangen 1980

Wolff, H. D.: Neurophysiologische Aspekte der manuellen Medizin, 2. Aufl. Springer, Berlin 1983